AF774218

Recent Results in Cancer Research 51

Fortschritte der Krebsforschung
Progrès dans les recherches sur le cancer

Edited by

V. G. Allfrey, New York · M. Allgöwer, Basel
K. H. Bauer, Heidelberg · I. Berenblum, Rehovoth
F. Bergel, Jersey · J. Bernard, Paris · W. Bernhard, Villejuif · N. N. Blokhin, Moskva · H. E. Bock, Tübingen · W. Braun, New Brunswick · P. Bucalossi, Milano · A. V. Chaklin, Moskva · M. Chorazy, Gliwice · G. J. Cunningham, Richmond · M. Dargent, Lyon · G. Della Porta, Milano · P. Denoix, Villejuif
R. Dulbecco, La Jolla · H. Eagle, New York · R. Eker, Oslo · R. A. Good, Minneapolis · P. Grabar, Paris
H. Hamperl, Bonn · R. J. C. Harris, Salisbury
E. Hecker, Heidelberg · R. Herbeuval, Nancy
J. Higginson, Lyon · W. C. Hueper, Fort Myers
H. Isliker, Lausanne · J. Kieler, København
G. Klein, Stockholm · H. Koprowski, Philadelphia
L. G. Koss, New York · G. Martz, Zürich
G. Mathé, Villejuif · O. Mühlbock, Amsterdam
W. Nakahara, Tokyo · L. J. Old, New York
V. R. Potter, Madison · A. B. Sabin, Rehovoth
L. Sachs, Rehovoth · E. A. Saxén, Helsinki
C. G. Schmidt, Essen · S. Spiegelman, New York
W. Szybalski, Madison · H. Tagnon, Bruxelles
R. M. Taylor, Toronto · A. Tissières, Genève
E. Uehlinger, Zürich · R. W. Wissler, Chicago

Editor in Chief: P. Rentchnick, Genève

Gliomas

Current Concepts in Biology, Diagnosis and Therapy

Edited by

J. Hekmatpanah

With 67 Figures

Springer-Verlag
Berlin · Heidelberg · New York 1975

Symposium on "Cerebral Gliomas" held at the University of Chicago, May 18—19, 1974

JAVAD HEKMATPANAH, M. D., Division of Neurological Surgery
The University of Chicago Hospitals, Chicago IL 60637/USA

Sponsored by the Swiss League against Cancer

ISBN-13:978-3-642-80918-7 e-ISBN-13:978-3-642-80916-3
DOI: 10.1007/978-3-642-80916-3

This work is subject to copyright. All rights are reserved, whether the whole or part of the material is concerned, specifically those of translation, reprinting, re-use of illustrations, broadcasting, reproduction by photocopying machine or similar means, and storage in data banks. Under § 54 of the German Copyright Law where copies are made for other than private use, a fee is payable to the publisher, the amount of the fee to be determined by agreement with the publisher.

© by Springer-Verlag Berlin · Heidelberg 1975.
Softcover reprint of the hardcover 1st edition 1975

Library of Congress Cataloging in Publication Data. Symposium on Cerebral Gliomas, University of Chicago, 1974. Gliomas. (Recent results in cancer research; v. 51.) "Sponsored by Swiss League against Cancer." Bibliography: p. Includes index. 1. Gliomas — Congresses. I. Hekmatpanah, Javad, 1934- ed. II. Schweizerische Nationalliga für Krebsbekämpfung und Krebsforschung. III. Title. IV. Series. [DNLM: 1. Brain neoplasms — Congresses. 2. Glioma — Congresses. W1 RE106P v. 51 / WL358 G559] RC261.R35 vol. 51 [RC280.B7] 616.9'94'008s [616.9'94'81]. 74-32388.

The use of registered names, trademarks, etc. in this publication does not imply, even in the absence of a specific statement, that such names are exempt from the relevant protective laws and regulations and therefore free for general use.

Preface

When I was asked by the Cancer Teaching Committee of the University of Chicago to set up a Symposium on Brain Tumors, I welcomed the opportunity to do so. But I felt that "Brain Tumors" was far too broad a subject to cover in a day and a half meeting. Furthermore, during the past decade, there have been a great many papers and symposiums on technical advancements in the treatment of benign brain tumors. On the other hand, while gliomas make up somewhere around 50% of all brain tumors, there have certainly been far fewer communications about them. For this reason I thought that it would be highly valuable to invite some of the leading investigators to share with us their experience with gliomas. The hope was to learn the current concepts about these tumors biologically and therapeutically, specifically to find out what we are doing and where we are going in these respects. Naturally all aspects could not be covered. Nevertheless, it was a widely expressed opinion by the speakers that they too learned while they came to teach. It is therefore hoped that the publication of their works and ideas through this monograph will also be useful for the reader. Most of the manuscripts were handed in at the time of the Symposium, and a few were sent later. Except for minor editorial changes the words are entirely those of the speakers and authors.
Dr. WISSLER introduced those who helped to set up the Symposium. To them, and to the speakers, I too extend my gratitude. I would like to thank Dr. WISSLER himself for initiating the concept and helping to make this and other Cancer Teaching Symposiums possible. Finally, I wish to thank EMILY SCHMIDT for her many hours of help in typing, editing, and proofreading.

JAVAD HEKMATPANAH, M. D.

Contents

Opening of the Symposium. R. W. WISSLER 1
Introductory Remarks. J. F. MULLAN 3

Session I
Chairman: J. F. MULLAN
Co-Chairman: N. A. VICK

The Significance of Experimental Gliomas for Human Disease. H. M. ZIMMERMAN 6
Refinement of the Avian Oncornavirus-Induced Primary Rat Brain Tumor Model for Therapeutic Screening. D. D. BIGNER, D. J. SELF, J. FREY, R. ISHIZAKI, A. J. LANGLOIS, and J. A. SWENBERG 20
Studies on the Differentiation of Human and Experimental Gliomas in Organ Culture Systems. L. J. RUBINSTEIN and M. M. HERMAN 35
Scanning Electron Microscopy of Cerebral Tumors and Glial Cells. A. ARNOLD and D. BURROWS 52
Discussion. N. A. VICK 63

Session II
Chairman: J. HEKMATPANAH
Co-Chairman: V. W. STEWARD

Neuroradiologic Aspects of Gliomas. P. WEINBERG 65
Computerized Tomography in Relation to the Diagnosis of Gliomas. M. S. HUCKMAN 79
Heavy Ion (Proton) Radiography in Medical Diagnosis. W. V. STEWARD 88
Treatment of Supratentorial Gliomas. P. C. BUCY 97
Treatment of Infratentorial Gliomas in Childhood. E. B. HENDRICK, H. J. HOFFMAN, and R. P. HUMPHREYS 102
Discussion. J. HEKMATPANAH 107

Session III

Chairman: M. L. Griem
Co-Chairman: D. H. Pearson

Thoughts on the Biology and Therapy of Malignant Gliomas.
W. M. Kirsch, J. J. van Buskirk, D. Schulz, and
K. Tabuchi 110

Chemotherapy: Current Results and Future Prospects.
C. B. Wilson 119

Conventional Radiation Therapy of Gliomas.
G. E. Sheline 125

Some New Radiotherapeutic Approaches and Combined Protocol Trials in the Management of Malignant Gliomas.
C. H. Chang 135

Discussion. M. L. Griem, D. H. Pearson 151

Closing Remarks. H. M. Zimmerman 153

Epilogue: Treatment of Glioma – How Far Have We Come, What Are We Doing, and Where Are We Going?
J. Hekmatpanah 159

List of Participants

ARNOLD, A., Hinsdale Medical Center and the Department of Surgery (Neurosurgery), The Pritzker School of Medicine, The University of Chicago, Chicago, IL

BIGNER, D. D., The Department of Pathology and Virology, Duke University Medical Center, Durham, NC

BUCY, P. C., Department of Surgery (Emeritus), Northwestern University Medical School, Chicago, IL, and Department of Neurology and Neurological Surgery, Bowman Gray School of Medicine, Wake Forest University, Winston-Salem, NC

CHANG, C. H., Department of Radiology, Columbia University, and Division of Radiotherapy, Columbia-Presbyterian Medical Center, New York, NY

GRIEM, M. L., Department of Radiology and The Chicago Tumor Institute, The Pritzker School of Medicine, The University of Chicago, Chicago, IL

HEKMATPANAH, J., Department of Surgery (Neurosurgery), The Pritzker School of Medicine, The University of Chicago, Chicago, IL

HENDRICK, E. B., Department of Neurosurgery, The Hospital for Sick Children, The University of Toronto, Toronto, Ontario, Canada

HUCKMAN, M. S., Department of Diagnostic Radiology, Presbyterian-St. Lukes Hospital and Rush Medical School, Chicago, IL

KIRSCH, W. M., Department of Neurosurgery, The University of Colorado Medical Center, Denver, CO

MULLAN, J. F., Department of Surgery (Neurosurgery), The Pritzker School of Medicine, The University of Chicago, Chicago, IL

PEARSON, D. H., Department of Surgery (Neurosurgery), The Pritzker School of Medicine, The University of Chicago, Chicago, IL

RUBINSTEIN, L. J., Department of Pathology (Neuropathology), Stanford University Medical Center, Stanford, CA

SHELINE, G. E., Department of Radiology (Radiation Oncology), The University of California School of Medicine, San Francisco, CA

STEWARD, V. W., Department of Pathology (Neuropathology), The Pritzker School of Medicine, The University of Chicago, Chicago, IL

VICK, N. A., Department of Medicine (Neurology), The Pritzker School of Medicine, The University of Chicago, Chicago, IL

WEINBERG, P. E., Department of Radiology (Neuroradiology), Northwestern Memorial Hospital, Northwestern University, Chicago, IL

WILSON, C. B., Department of Neurosurgery, The University of California Medical Center, San Francisco, CA

WISSLER, R. W., Department of Pathology, The Pritzker School of Medicine, The University of Chicago, Chicago, IL

ZIMMERMAN, H. M., Department of Pathology, Montefiore Hospital and Medical Center, and the Albert Einstein College of Medicine, New York, NY

Opening of the Symposium

R. W. Wissler

This is the 6th biennial cancer teaching symposium that we have organized and held at the University of Chicago under the auspices of the Clinical Cancer Training Program which is supported by a grant from the National Cancer Institute. Four of these have been published previously as a part of this Springer-Verlag series, Recent Results in Cancer Research. This is the fifth one.

The main purpose of these teaching efforts has been to bring to our house staff, students and young faculty and to many guests from Chicago and more distant centers the most up-to-date information on various aspects of a significant subject in the field of clinical neoplasia. We have attempted to present a coordinated overview of that subject, something that it is difficult to do on most medical school campuses with the locally available faculty.

Now to achieve this goal of a modern, up-to-date overview of an important subject in human cancer and to make it worth publishing, one has to have several ingredients:

1. One needs to have adequate financial support, in this case provided by 5 T12 CA-08077.
2. The subject needs to be important and timely. To choose the subject for these teaching symposia, including this one, we have had the constant help of a very talented advisory committee for the Cancer Training Program. They have spent a lot of time and effort on these decisions. They include: GEORGE BLOCK, MELVIN GRIEM, ELWOOD JENSEN, WERNER KIRSTEN, JOHN MULLAN, CHARLES PLATZ, HENRY RAPPAPORT, JOHN ULTMANN and GEORGE WIED.
3. One must have a knowledgeable, hardworking program chairman who should be an outstanding teacher. In this case Professor Dr. JAVAD HEKMATPANAH has devoted an immense amount of time and effort in organizing an excellent program, with well selected topics, and most important of all, superior teachers.
4. The faculty for the teaching symposium is the fourth ingredient. There must be an outstanding faculty and a thoroughly competent and experienced summarizer or rapporteur who has great skill. Both the faculty and the person chosen to summarize the conference must be outstanding. You can judge how well we have done but I believe you will enjoy great teaching as you read this volume.
   There is a spin-off from this, of course, because the faculty at each previous symposium has consistently told us that they also have learned a great deal!
   We can only express our very great appreciation for the efforts of the faculty, the time devoted to this symposium and the manuscripts and we hope that these illustrious teachers also found their participation in this symposium rewarding.
5. The fifth necessary component of any good teaching symposium is that small group of more or less anonymous people who help behind the

scenes - in this case JULIE KANT is the key person who has in her highly intelligent, cheerful and energetic way worked with EMILY SCHMIDT and LAURA STEELE to make all the arrangements for the meeting. She has, in addition, helped in many ways to bring the manuscript to fruition.

6. Finally we must acknowledge the good audience of interested students, many of whom were already experts in the neurological sciences and some of whom came from great distances to attend.
   We think that the people who really wanted to learn about gliomas were there and we hope that there will be an ever increasing circle of interested students who will profit from our efforts to produce both a learned and learning monograph on gliomas.

Introductory Remarks

J. F. MULLAN

I would like to join with my colleagues in welcoming all of you here. As you can gather, one of the purposes of this meeting is to spare our students and residents from the labors of listening to the staff; they have listened to us for years. Consequently our participation is kept to a minimum, so I will try to make this a minimal introduction. As Dr. WISSLER mentioned, there are really two purposes in this type of meeting. One is the factual knowledge gained. Another is perhaps more important. It is that this meeting presents an opportunity for our younger people to meet the leaders in the field. All of us, as we have settled upon our careers, have been influenced somewhere or other along the line in that decision. It may have been that the intellectual content of our specialty drove us to it, but I think in many cases it has not been the intellectual content, but the contact with an individual or individuals who stimulated our interests and our enthusiasm for this particular line of work. That, I think, is probably the most important ingredient of today's meeting, because we have with us today, without any question, a top talent in this field.

For a long time the whole problem of brain tumors was in the hands of the surgeons, since GODLEE took the first tumor out in 1884, 90 years ago. Since then, neurosurgeons have been mainly responsible for tumors. They have been responsible for surgery, and they have been responsible to a large extent for pathology, particularly the founder of this department of neurosurgery, PERCIVAL BAILEY, who evolved the classic descriptive pathology of brain tumors. In contrast studies too, surgeons took a leading role in all these decades from DANDY'S introduction of air studies to the introduction of technetium scanning by PAUL HARPER, who is a general surgeon at this University. But just as it is not safe to leave wars to the generals, it was realized long ago, it is not quite safe to leave brain tumors to the surgeons. Some of the more astute pathologists many years ago recognized this deficiency, and we delight in welcoming Dr. ZIMMERMAN, who was responsible for really turning descriptive neuropathology of brain tumors into experimental neuropathology. That was a most significant development. We also welcome Dr. RUBINSTEIN, who now represents the second generation of neuropathologists dedicated to the study of brain tumors. There has been a long and close relationship between surgery and pathology. We have many individuals who did not quite know which discipline to pursue. Dr. FERNANDEZ-MORAN of this University, who unfortunately is not here today, was doing a neurosurgical residency in Stockholm when Dr. OLIVECRONA took him aside, saying "You've got to be either a basic scientist or neurosurgeon; what do you really want to do?". Dr. FERNANDEZ-MORAN chose basic studies, and he has been largely responsible for the introduction of the electron microscope and the development of its techniques into the study of neuropathology. He still feels himself something of a surgeon, and I think whichever way his career would have gone would have been a gain to that area. This dilemma persists. I believe that Dr. BIGNER started off to be a neurosurgeon, but he too sidetracked somewhere along the way. And Dr. ARNOLD never really was sure which he was; he moved from surgery to pathology and back again. He was the individual who first observed the amytrophic lateral sclerosis in Guam because his vision encompassed more than the narrow range of technical neurosurgery. Currently he is back again into pathology, probing the uses of the scanning microscope. I think that during our training the designated direction of our career specialty should not outweigh all other considerations. I believe we should use all the neurosciences together to tackle any particular problem, whether it is a tumor, or multiple sclerosis, or a vascular problem. Whatever it is, we should not feel narrowly regimented by our training experience. I believe we surgeons are currently in danger of becoming less productive because we are too narrow. But for a long time neurosurgery was well represented, not only in pathology, but in other fields too. If PERCIVAL BAILEY was the surgeon's pathologist, I guess WILDER PENFIELD was the surgeon's physiologist. There were, however, several areas left unrepresented. We didn't have the surgeon's biochemist until WOLFF KIRSCH came along, and I believe now we can say we have one. We haven't had a clinical pharmacologist until CHARLIE WILSON and others of his generation came along. Just as we talk about the giants of 30 or 40 years ago, such men as BAILEY, DANDY, and PENFIELD, I believe 30 or 40 years from now we will recall today's guest faculty of biochemists, virologists, pathologists, radiologists, and clinical pharmacologists with corresponding indebtedness. As I see the future of neurosurgery, it doesn't consist of bigger and better surgery, it consists of more broadly trained neurosurgeons who are competent in the basic sciences, as competent in the basic sciences as they are in neurosurgery. Neurosurgery itself has probably come somewhere near the end of its technical evolution. An example of this: our chairman, Dr. HEKMATPANAH, finished his training in 1963, and as far as I know has not yet lost a patient in relation to the surgical treatment of brain tumor, either as a resident or subsequently. As far as surgical mortality goes, we have arrived; we cannot improve upon his record. Just as he is an example of a current product of a modern training program under JOSEPH EVANS, I think we would also like to welcome back the first neurosurgical graduate of the training program at the University of Chicago, Dr. PAUL BUCY -- Dr. PAUL BUCY of BAILEY, BUCY and BUCHANAN's textbook, of the Klüver and Bucy syndrome, BUCY the outstanding surgeon of the thirties, of the forties, of the fifties, of the sixties, and of the seventies, BUCY the founder of Surgical Neurology, the individual who has been mainly instrumental in getting the National spinal cord program on the way. Looking back, PAUL, we feel your spirit never really left us. I also welcome Dr. BRUCE HENDRICK of that exacting neurosurgical school of Toronto. I think that where clinical judgement and surgical skill reign supreme, the successors of MCKENZIE and BOTTERELL must always have a place upon the podium. Though the technical evolution of surgery is somewhere near the end, the role of neurosurgery is perhaps not yet defined, because, theoretically at least, if one can detect a tumor early enough, one can take it out entirely. Our problem has been getting at it early enough. And so we look forward to our colleagues from our sister Universities of this city, to Dr. PETER WEINBERG and Dr. HUCKMAN, to tell us where neuroradiology is going, because if we can get those tumors earlier, we can cure them; it depends upon how early. Dr. WEINBERG will discuss conventional methods while Dr. HUCKMAN will introduce us to computerized axial tomography. The rules of this meeting prevent us hearing from Dr. STEWARD, our Co-Chairman of the afternoon, who is working on the proton beam scanner; it's an interesting device. By focusing the Bragg peak on the receptor system rather than on the tumor itself, one can get an enhanced contrast with minimal radiation exposure. So one can at least theoretically envision at this moment getting a proton beam as simply and with as little radiation as one can get a routine x-ray of the chest now. If we can get a routine early radiological 3-dimensional view of the brain without invasive contrast, then surgery will have gained a new role in brain tumors. There are other areas that must be explored at the present time, and which have nearly reached the end of their potential, especially radiotherapy, so I look forward very much to hearing what Dr. SHELINE and Dr. CHANG have to say there. We have a conventional system of delivering radiotherapy five days a week. Maybe if we give a little radiotherapy every day for a year, it might be better. Maybe if we did this like LEKSELL, using stereotactic methods of focused beams, we would do better. Maybe, as Dr. GRIEM has suggested, we must simply change the cerebral metabolism. Maybe we should make more extensive use of the radio-protective drugs, the radio-enhancement drugs. This still is quite a wide field. We have not included in this program any major contribution on immunology, and the reason is, I think, this is not far enough along for proper evaluation. But as one aspect of neurosurgery has reached the limit, all these other aspects of investigation and treatment of brain tumors are on the rise. And certainly our position today will look as outdated 30 years from now as does the position of those who were working at the early, exclusively neurosurgical, phase. So I look forward with great enthusiasm to this symposium, and again would like to thank all the speakers who have come. I like to again remind the students and those early in their careers that you are here for these two days to hear the best that is going on in this area, from masters in their fields. And I think, for those of you who are undecided in your careers, it might be not so much in what these people say as in the satisfaction and enthusiasm in the investigative life, that you might find information and experience to your profit.

The Significance of Experimental Gliomas for Human Disease

H. M. ZIMMERMAN

EPIDEMIOLOGY OF HUMAN INTRACRANIAL NEOPLASMS

In 1962 KURLAND and his associates (1) compiled the available mortality data on cerebral neoplasms for the United States and Canada as derived from the International List of Diseases and Causes of Death (2). Their study revealed a remarkably even geographic distribution for all classes of intracranial tumors in the two countries. The mortality rates ranged from 3 to 5 per 100,000 population. If this mortality estimate was correct in 1962 and if it is still valid in 1974, one can expect between 6,300 and 10,500 deaths from intracranial neoplastic disease this year in this country based on a population of two hundred and ten million.

The compilation of data of this kind is subject to innumerable inaccuracies, as pointed out by KURLAND and his associates. They noted, for example, that the rate for deaths due to brain tumors in Japan was very low, being about half that in Western countries, and they felt that at least in part this low rate could be attributed to diagnostic and reporting artifacts.

These investigators further reported the results of a 10 year survey among the population of Rochester, Minnesota. They found a prevalence ratio of intracranial neoplasms of 46 per 100,000 population and an incidence rate of about 19 per 100,000 per year. In this population, about 1 per cent of all deaths was due to primary intracranial tumors, which was about twice that reported in official mortality statistics.

Other studies have been carried out in the United States that disclose rates that vary from 3.0 - 8.4/100,000 of the population (3, 4, 5), and in other parts of the world the incidence also varies considerably (6, 7). Based upon confirmed tumor diagnosis at biopsy or autopsy, it has been estimated that primary intracranial neoplasms constitute 2-3 per cent of all neoplasms to which man is heir (8). There exist published accounts of the incidence of various expanding, space-occupying intracranial lesions, not all new-growths and not all primary neoplasms of the brain, that have been collected in different countries of the world. Some of these statistical reports have been presented in table form by this author (3), revealing wide discrepancies in the incidence of certain categories of tumors from country to country and even from series to series of cases collected by different authors in the same country.

Despite the reported variations in incidence of intracranial tumors in different series, all contributors agree that the gliomas rank first on a numerical basis. They constitute between 31 to 43 per cent of all intracranial masses, including metastatic tumors and granulomas such as tuberculomas. Simply eliminating the latter two categories of expanding intracranial lesions from consideration has the effect of raising the glioma incidence to between 40 and 50 per cent. Thus it can be seen that for numerical considerations alone this class of brain tumors is the most important (9). But there are many additional factors, some subject to experimental probing, as will be discussed below, that make it so important.

Among the human gliomas, this author has found that the relatively malignant astrocytic tumor of the glioblastoma multiforme variety occupies first place numerically with an incidence of over 51 per cent (3). Next in frequency is the more slowly growing, hence more benign, astrocytoma (incidence of 24.5 per cent). The ependymoma is next with an incidence of slightly more than 6 per cent, followed by the oligodendroglioma (5.5 per cent). The polar spongioblastoma, confined essentially to the pons and brain stem and occasionally to the corpus callosum, was present in 3.4 per cent of the 1,633 glioma tumors of this series. Attention is drawn to the fact that 57 (3.4 per cent) of all the gliomas were mixed tumors; i.e., they contained more than one identifiable gliogenous component. Among these "mixed" gliomas were some that were composed of both ependymomatous and oligodendrogliomatous parts, and some that had additional astrocytomatous as well as sarcomatous portions.

THE EXPERIMENTALLY PRODUCED GLIOMAS

It is fortunate that the experimentally produced gliomas often resemble morphologically their human counterparts so closely as to be virtually indistinguishable from them. Thus the cerebral astrocytoma, produced in mice by intracerebral implantation of compact pellets of chemical carcinogens such as methylcholanthrene, dibenzanthracene or benzpyrene (10), is a microcystic tumor composed of stellate astrocytes (Fig. 1) very much as are the human astrocytomas. In general, the more slowly growing tumors of astrocytic origin are produced less frequently with the aromatic hydrocarbons than are the more malignant neoplasms.

By far the most frequent murine glioma produced experimentally is the glioblastoma multiforme (Types III and IV astrocytoma). It is this tumor, as already mentioned, that occupies first place among the human gliomas. Like the latter, the murine neoplasm is highly pleomorphic, with foci of necrosis and marginal spongioblasts usually arranged in palisades (Fig. 2). Multinucleated tumor giant cells are common. Vascular thromboses and hemorrhages are also frequent features of this neoplasm.

A modification of the murine glioblastoma multiforme exists in which the neoplasm, in addition to disclosing all the cytologic details just enumerated, is also characterized by numerous huge multinucleated neoplastic cells (Fig. 3). This is the tumor which in man is sometimes called "gigantocellular" or "monstrocellular" glioblastoma multiforme. The suggestion that the human tumor may be of mesodermal, hence sarcomatous, origin is not valid for this tumour in the mouse, which is unequivocally a glioma as shown by special staining methods.

Still another variation of the malignant form of glioma of the astrocytic variety seen in the corpus callosum, pons or brain stem of the

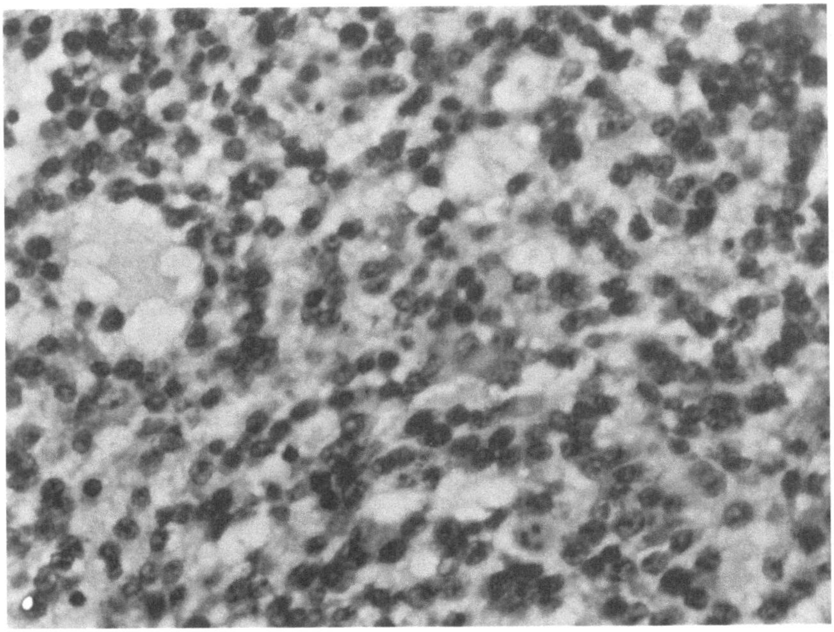

Fig. 1. Microcystic cerebral astrocytoma. x. 425*

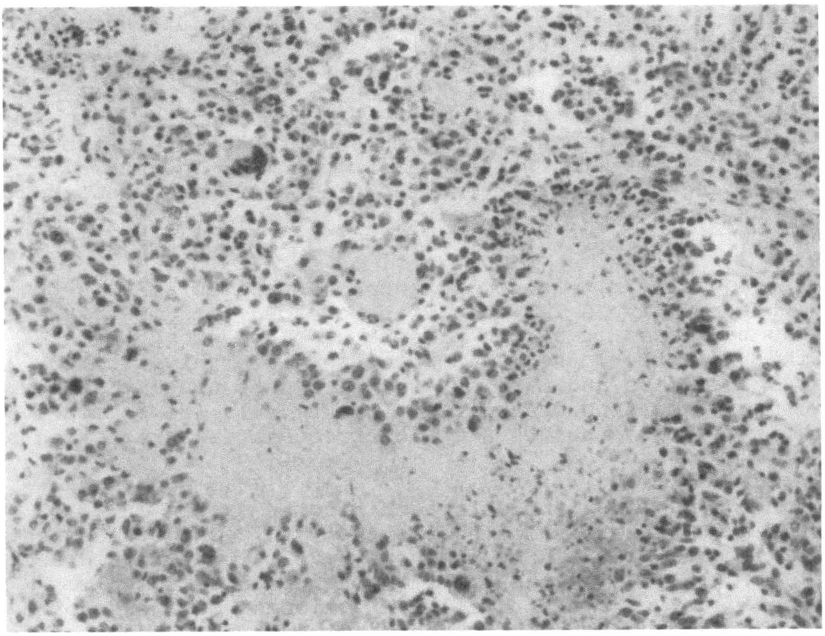

Fig. 2. Glioblastoma multiforme. The acellular area consists of necrotic tissue. The tumor cells are pleomorphic and a multinucleated tumor giant cell is present. x 180

* The first eleven figures are of tumors embedded in paraffin and stained with hematoxylin-eosin.

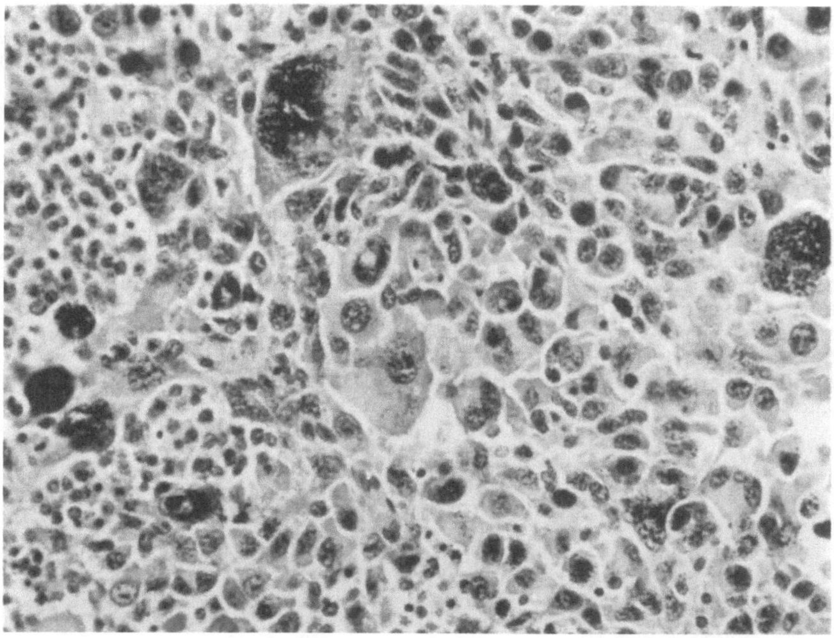

Fig. 3. Gigantocellular glioblastoma multiforme. x 275

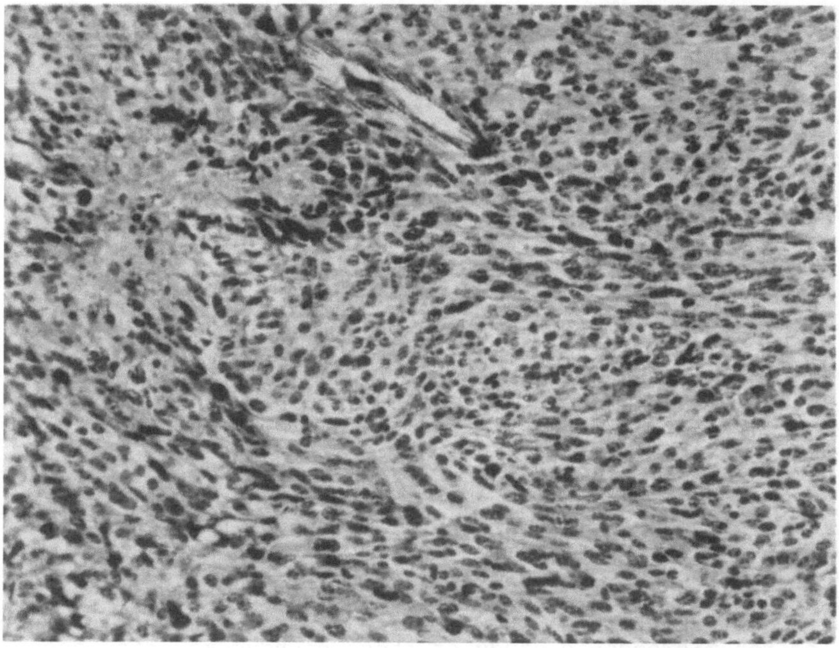

Fig. 4. Polar spongioblastoma in corpus callosum. Note interlacing bundles of spongioblasts. x 275

mouse is characterized by large interlacing bundles of bipolar and unipolar spongioblasts (Fig. 4). There are present occasional necrotic foci in some of these tumors and also a few giant cells with multiple nuclei. A considerable degree of cellular pleomorphism is always present. In man, this tumor is designated by the term "polar spongioblastoma."

With pellets of carcinogen implanted in the subcortical white matter of the animal brain, there is often produced a glioma of distinctive cellular appearance. The rather large uniform tumor cells are compartmentalized into lobules, almost gland-like in appearance, by vascularized connective tissue septa (Fig. 5). The individual neoplastic cells have clear cytoplasm between the chromatin-rich, round nuclei and the cellular membranes. The pale cytoplasm creates the appearance of "haloes" around the nuclei. This tumor is the easily recognized oligodendroglioma. Less frequently than the similar tumor in man, this murine glioma has calcium salt deposits.

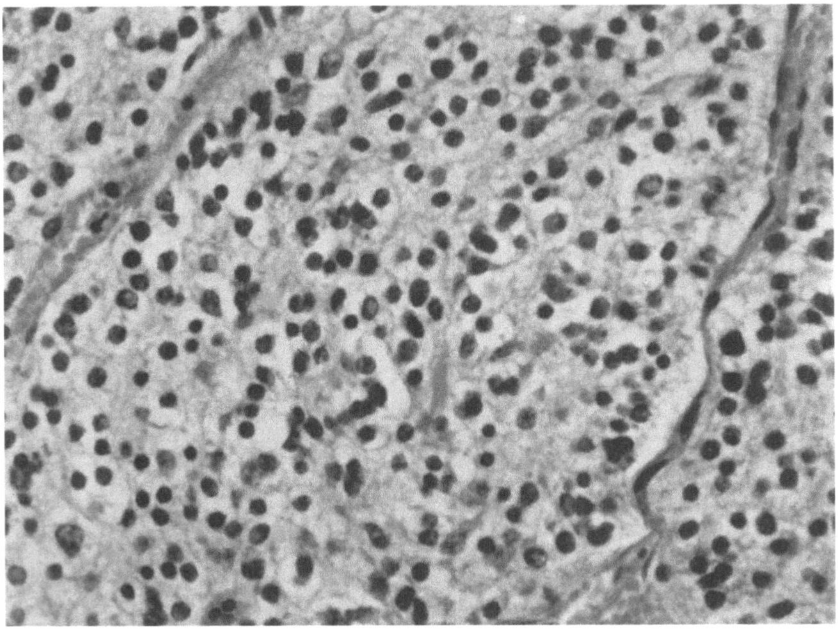

Fig. 5. Oligodendroglioma. The characteristic cells have round nuclei with clear cytoplasm that imparts an appearance of perinuclear haloes. Vascularized connective tissue septa divide the tumor into lobules. x 450

When the chemical carcinogen is implanted intraventricularly, there very often appears a tumor whose origin can be traced to the lining ependymal cells. Characteristically, the tumor cells form rosettes either around blood vessels or around acellular centers that are filled with delicate fibrils whose origin is from the neoplastic cells (Fig. 6). Electron microscopic study of these ependymomas discloses the same junctional complexes and cilia (or their precursors known as "basal bodies") that are diagnostic of the human tumor of this variety (11, 12).

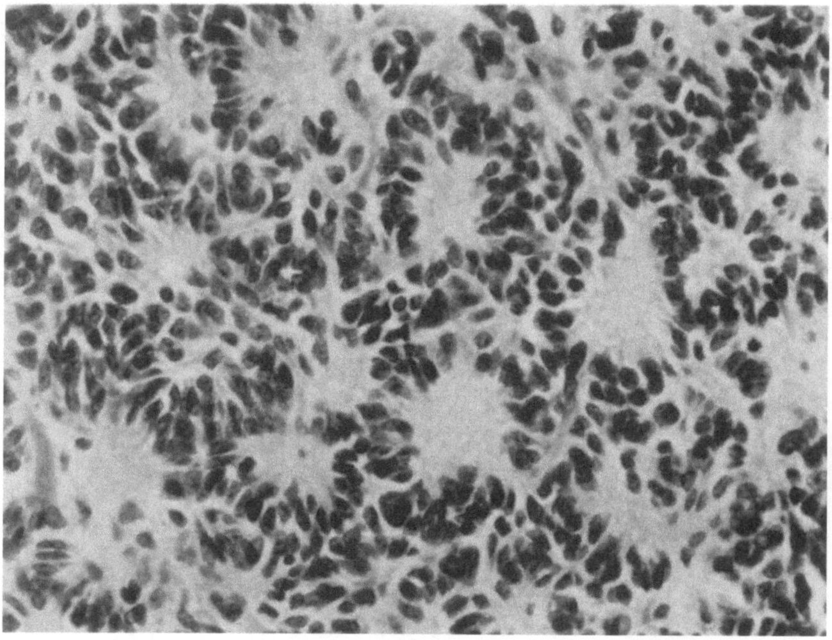

Fig. 6. Ependymoma. Tumor cells form rosettes around centers of glial fibrils. x 450

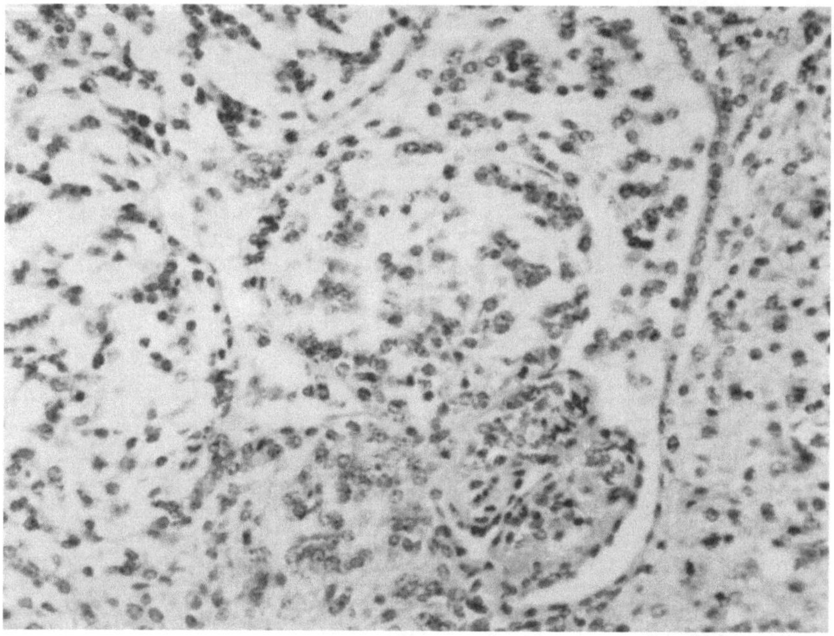

Fig. 7. Myxopapillary ependymoma. The tumor is seen to take origin from the ependyma lining the ventricle. x 275

There is a variation of this tumor in man that is called a myxopapillary ependymoma, and precisely this morphologic variant is sometimes seen in the experimental animal (Fig. 7). Its origin from the ventricular ependyma is clear. It is composed of tumor cells arranged in papillae between which is deposited a proteinaceous, myxomatous, PAS-positive substance.

Still another variation of the ependymoma is seen in the experimental animal and is similar to the rare human medulloepithelioma (13). This neoplasm has a tubular and papillary arrangement of its columnar cells (Fig. 8). In part it appears to be well differentiated, but some portions of the tumor are quite primitive in appearance and contain numerous cells in mitotic division.

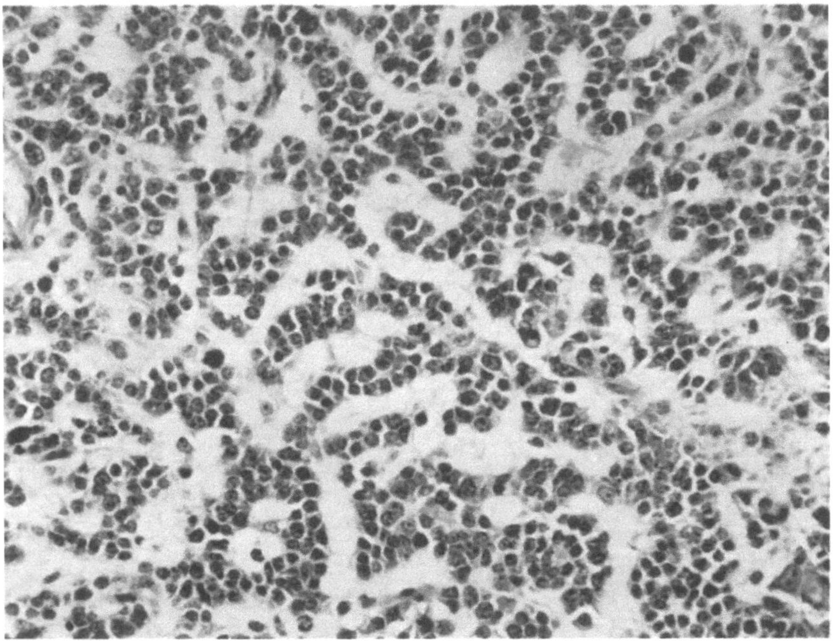

Fig. 8. Medulloepithelioma. The columnar neoplastic cells form tubules and papillary structures. x 275

CYTOGENESIS AND ETIOLOGY OF EXPERIMENTAL GLIOMAS

The murine gliomas induced with hydrocarbons are derivatives of the three basic glial cell types of the adult animal brain: the astrocyte, the oligodendrocyte and the ependyma. It is emphasized that in 35 years of experimentation during which several thousand mouse brain tumors were produced, not a single neoplasm developed whose origin could be traced to the fourth glial element of the brain; namely, the microglial cell. It is further emphasized that all the gliomas were invariably induced in adult animals and that the tumorigenic process was started by chemical implantation in the brains of young adult animals (from 6 weeks to 6 months of age). This was done for the purpose of reasonably excluding the possibility of the interaction of the carcinogens with immature or embryonic glia, or embryonic cell rests. It may account for the fact that not a single ganglioglioma was ever produced, presumably because after neurons reach maturity and are fully differentiated they no longer are capable of neoplastic transformation. In a few instances when tumors developed in the cortical gray matter or the basal ganglions, a superficial appearance of gangliogliomas resulted (Fig. 9). The malignant glial cells in such tumors were easily recognizable and mature neurons were found among them, but the neurocytes were not neoplastic. They appeared merely to have been entrapped by the proliferating glia.

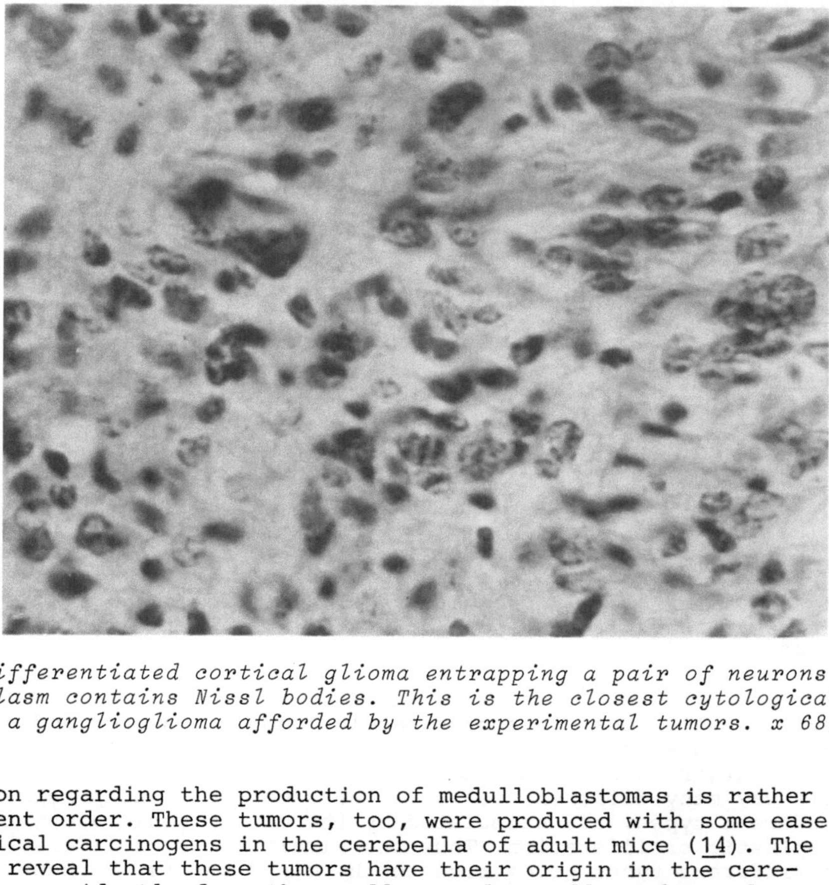

Fig. 9. Undifferentiated cortical glioma entrapping a pair of neurons whose cytoplasm contains Nissl bodies. This is the closest cytological approach to a ganglioglioma afforded by the experimental tumors. x 680

The situation regarding the production of medulloblastomas is rather of a different order. These tumors, too, were produced with some ease and by chemical carcinogens in the cerebella of adult mice (14). The experiments reveal that these tumors have their origin in the cerebellar cortex, evidently from the small granular cells and not from glia (Figs. 10, 11). This would make the medulloblastoma a neuroblastoma, and not a glioma at all. It would seem that in the process of neuronal differentiation in the cerebellar folia the small granular cells had not achieved the degree of maturation as to preclude neoplastic transformation, as is the case with the larger ganglion cells of the cerebellar cortex.

The problem of the "mixed" gliomas, as well as of the mixed gliomasarcomas, seen both in man and the mouse seems to be readily explained on the basis of the following consideration. The small carcinogenic

Fig. 10. Medulloblastoma. The tumor cells occupy the inner granular cell whose neurocytes they closely resemble. At the top of the figure, the acellular cleft represents the interfoliar subarachnoid space. Note that there is no outer granular cell layer in this adult mouse. x 140

pellets, which measure approximately 1 mm in diameter and are implanted in the brains for the purpose of producing the neoplasms, are many times larger than any of the cells in whose midst they are placed. In this way many different cells of the same type, and possibly also of different types, are stimulated by the carcinogenic agent to undergo neoplastic proliferation. Thus the simultaneous proliferation of cells of different types gives rise to the mixed tumors. Perhaps the marvel is that such tumors do not actually occur more frequently than they do, for "pure" tumors, i.e. those of one cell type, can evidently occur if only a single cell gives rise to the neoplasm or if only cells of the same and not other types are affected.

A more detailed study of the mechanism of carcinogenesis with chemical agents than was afforded by light microscopy was made possible with the availability of the electron microscope. It had been shown that the induction of cerebral neoplasms with carcinogens followed the process of phagocytosis of the chemical crystals by those cells destined to become neoplastic (15). What ensued within the chemical-bear-

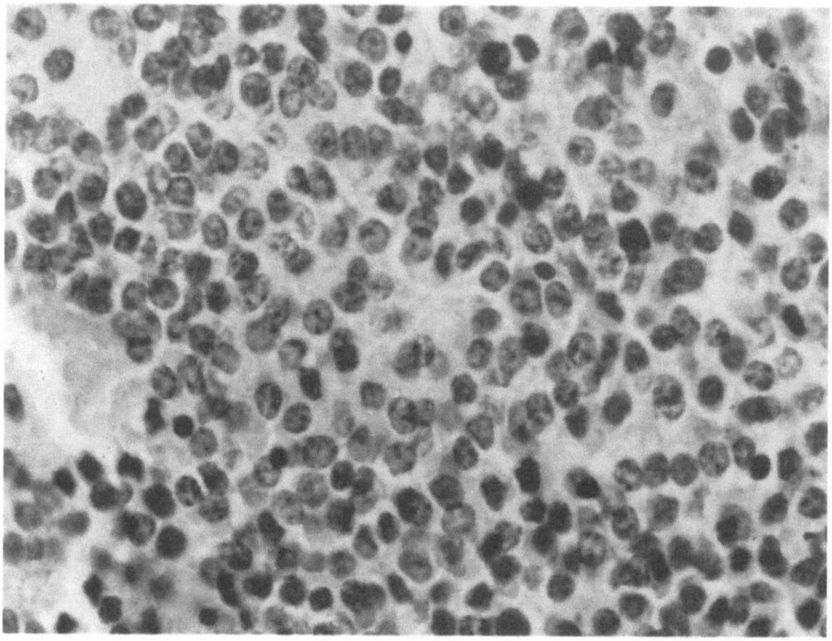

Fig. 11. Medulloblastoma. Higher magnification of tumor cells in figure 10. The neuroblasts ("medulloblasts") are uniform in appearance and there is practically no stroma, either glial or mesodermal. In the very center of the figure, some tumor cells are arranged so as to suggest a rosette. x 720

ing cells that resulted in their neoplastic transformation was suggested by several electron microscopic studies (16, 17, 18). Within the cytoplasm of the cells reacting to the carcinogenic crystals there appeared uniform filamentous and spherical particles with an outer diameter of about 77 m^{μ} (Fig. 12). These intracellular particles were present only during the precancerous stages of the cellular response to the pellet implantation; they disappeared from the cells when a glioma could first be recognized microscopically. Efforts to produce tumors with these virions in saline suspensions that were free of whole tumor cells and of carcinogen resulted in the production of malignant neoplasms at the injection sites in 8 of 14 animals (8). Additional confirmatory evidence is still necessary for the establishment of these virions as agents in experimental gliomatogenesis. No evidence of any kind is as yet available that would implicate viruses as etiologic agents in human gliomas.

From time to time it was suggested in the past that the human glioblastoma multiforme was caused by an infectious agent (19). It was the following considerations that were responsible for this view: 1, this tumor frequently has extensive necrosis; 2, it often has a partial mesodermal stroma; 3, there are vascular proliferative changes suggestive of granulation tissue; 4, there are vascular thrombotic occlusions; 5, lymphocytic accumulations may surround some blood vessels; and 6, some multinucleated giant cells may resemble those seen in granulomas. Of course, none of these features prove an infectious etiology -

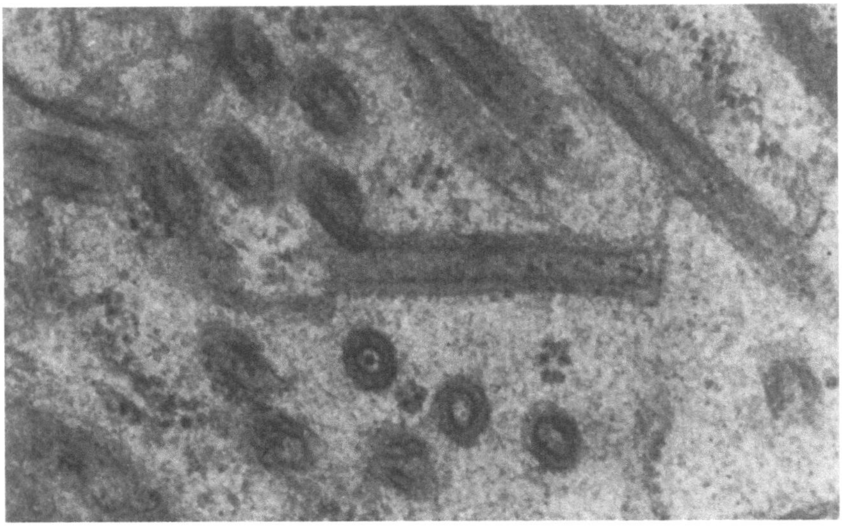

Fig. 12. Intracytoplasmic filamentous and spherical virions in a precancerous lesion produced with dibenzanthracene in the brain of an adult mouse. Electron micrograph x 100,000

all early attempts to culture significant organisms in these tumors failed. In more recent times, it was demonstrated that some nuclei in the tumor cells contained structures that were somewhat suggestive of viral inclusion bodies. But electron microscopy proved conclusively that these intranuclear bodies were merely cytoplasmic invaginations.

Other chemical agents in addition to the aromatic hydrocarbons have been employed in the production of experimental brain tumors. The resorptive carcinogens such as methylnitrosourea (MNU) and ethylnitrosourea (ENU) have the advantage of employability in the rat, a larger animal than the mouse, and can be fed or injected intravenously (20, 21). Employed in the pregnant rat, ENU passes through the placenta and produces neural tumors in the offspring (22). These tumors are frequently multiple and their locations in the nervous system are quite unpredictable; among favorite sites are the Gasserian ganglion and the spinal cord. Of the gliomas produced with these compounds, there seem to be high incidences of oligodendrogliomas and mixed gliomas. Of the nonglial tumors, both intra- and extracranial neurilemomas are prominent.

Beginning in the mid-1930s work was started with viral agents in attempts to produce brain tumors in animals. First, the Rous sarcoma virus (RSV) was injected in chicks with some success and later, in the 1960s, the same agent was employed in hamsters, rabbits and dogs. The astrocytoma was the most common glioma produced in mammals with the RSV, but spongioblastomas and glioblastomas multiforme were not uncommon. Soon other viruses were put into use in brain tumor production by an increasing number of investigators. The results of these experiments were well summarized by IKUTA and KUMANISHI (23), who reported significant results of their own with the human adenovirus type 12 in hamsters. That the human adenovirus type 12 is oncogenic for animals is of considerable importance and has already stimulated intensive search for other human oncogenic viruses. Most recently MUKAI and KOBAYASHI (24) described the results in newborn hamsters that received intracerebral inoculations of the human adenovirus. Their animals developed cerebral and spinal tumors of remarkable cellular uniformity that with light and electron microscopy were classified as ependymomas. It is this glioma that other investigators also have reported when employing adenovirus type 12.

IN EXPLANATION OF CERTAIN GLIOMA PROBLEMS

In 1957 this writer touched on the significance of experimental gliomatogenesis in explaining some problems with human gliomas (25). The more important of these problems are the following:

1. Differentiation and Dedifferentiation of Tumor Cells. It has long been established that the cells of the nervous system including glia reach maturity by a process of differentiation morphologically and functionally. In neoplasia, it has been assumed that a process of dedifferentiation accounts for the anaplastic cells associated with malignancy. As part of this concept it has also been assumed that in the formation of a glioma a single differentiated cell is somehow stimulated to proliferation and that the resulting neoplasm has its origin from that one cell. The cellular pleomorphism of any glioma is attributed, on the basis of this theory of the unicellular origin of tumors, to dedifferentiation. But it has already been shown in the discussion of the origin of mixed gliomas in animals that the diverse cellular composition of these tumors is a function of multicellular stimulation to proliferation. Dedifferentiation undoubtedly accounts for the presence of astroblasts and spongioblasts in a tumor of astrocytic origin, but it cannot conceivably account for the presence of astrocytes in an ependymoma, or oligodendroglioma, or medulloblastoma, nor of ependymal cells in an oligodendroglioma. The concept of dedifferentiation as the explanation for pleomorphism in human gliomas therefore has restricted applicability. Like the extracranial neoplasms of the malignant lymphoma group that have a multicellular origin, this is also the case of many gliomas, both human and experimental.

2. Extracranial Metastasis of Gliomas. It is a well known fact that spontaneous extracranial metastasis of human gliomas practically never occurs, even with the most malignant of the tumors such as the glioblastoma multiforme. This is also true for the experimental neoplasms. From time to time it has been reported that small numbers of tumor cells have been found in the blood aspirated from the superior longitudinal sinus of patients with gliomas, but these tumor cells evidently fail to survive in sufficient numbers to provide active growth in other organs of the body. When gliomas in man have been found in metastatic foci extracranially, they have occurred almost without exception following surgery, when it was assumed that inadvertent transplants of tumor cells occurred intravascularly. This writer has seen metastatic human gliomas in the lungs and liver following surgery. He has also seen such tumors lining the right pleural and the peritoneal cavities in patients who have had ventriculopleural and ventriculoperitoneal shunts for obstructive hydrocephalus due to tumors.

The experimental glioma studies have helped validate the clinical assumptions regarding extracranial metastases. First, it has been shown that experimental tumors of this class are incapable of invading cerebral blood vessel walls, unlike primary intracranial mesodermal tumors, and hence they are sealed off from effective pathways of communication between the brain and the other organs of the body. Then it was shown in numerous experiments (26, 27) that such primary experimental gliomas when provided with an egress from the skull will grow with ease in extra-cranial locations. Such egress is provided by the mechanical transplantation of fragments of glioma with a trochar in the subcutaneous tissues of the flank in homozygous mice. Or they will grow readily in the anterior chamber of the eye in mice and other animals. In experiments performed in my laboratory by Dr. N. KAGEYAMA some years ago, it was shown that saline suspensions of tumor homogenates when injected into the jugular veins of mice produced pulmonary gliomatous metastases with great ease. Likewise, such suspensions injected into the internal carotid artery resulted in intracerebral metastases. These experiments proved conclusively that gliomas will grow extracranially if only they are provided the necessary transport system.

REFERENCES

1. KURLAND, L. T., MYRIANTHOPOULOS, N. C., LESSELL, S.: Epidemiologic and genetic considerations of intracranial neoplasms. In: The Biology and Treatment of Intracranial Tumors, p. 5. FIELDS, W. S. and SHARKEY, P. C. (eds.). Springfield, Ill.: Charles C. Thomas 1962.
2. Manual of the International Classification of Diseases, Injuries, and Causes of Death: Sixth Revision, Vol. 1. Geneve: World Health Organization 1948.
3. ZIMMERMAN, H. M.: Introduction to tumors of the central nervous system. In: Pathology of the Nervous System, Vol. 2, p. 1951. MINCKLER, J. (ed.). New York: McGraw-Hill 1971.
4. DORN, H. F., CUTLER, S. J.: Morbidity from Cancer in the United States. Public Health Monograph 29, U. S. Dept. of Health, Education and Welfare, Public Health Service 1955.
5. HAENSZEL, W., MARCUS, S. C., ZIMMERER, E. G.: Cancer morbidity in urban and rural Iowa. Public Health Monograph 37, U. S. Dept. of Health, Education and Welfare, Public Health Service 1956.
6. CHEN, K.-M., BRODY, J. A., KURLAND, L. T.: Patterns of neurologic diseases on Guam. I. Epidemiologic aspects. Arch. Neurol. 19, 573 (1968).
7. COHEN, A., MODAN, B.: Some epidemiologic aspects of neoplastic diseases in Israeli immigrant population. III. Brain Tumors. Cancer 22, 1323 (1968).
8. ZIMMERMAN, H. M.: Brain Tumors: Their incidence and classification in man and their experimental production. Ann. N. Y. Acad. Sci. 159, 337 (1969).
9. ZIMMERMAN, H. M.: The ten most common types of brain tumor. Seminars Roentgenol. 6, 48 (1971).
10. ZIMMERMAN, H. M.: Brain tumors. In: Methods in Cancer Research, Vol. 10, p. 105. BUSCH, H. (ed.) New York: Academic Press 1973.
11. POON, T. P., HIRANO, A., ZIMMERMAN, H. M.: Electron Microscopic Atlas of Brain Tumors. New York: Grune & Stratton 1971.
12. HIRANO, A., GHATAK, N. R., ZIMMERMAN, H. M.: The fine structure of ependymoblastoma. J. Neuropathol. Exptl. Neurol. 32, 144 (1973).
13. JELLINGER, K.: Cerebral medulloepithelioma. Acta neuropath. 22, 95 (1972).
14. ZIMMERMAN, H. M.: The histopathology of experimental "medulloblastoma." Acta neuropath. 8, 69 (1967).
15. ZIMMERMAN, H. M.: The nature of gliomas as revealed by animal experimentation. Am. J. Pathol. 31, 1 (1955).
16. IKUTA, F., ZIMMERMAN, H. M.: Virus particles in reactive cells induced by intracerebral implantation of dibenzanthrene. J. Neuropathol. Exptl. Neurol. 24, 225 (1965).

17. POPOFF, N., SUTTON, C. H., ZIMMERMAN, H. M.: Virus-like particles in reactive cells associated with crystals of implanted carcinogen. Acta neuropath. 10, 308 (1968).
18. IKUTA, F., ZIMMERMAN, H. M.: Virus particles induced with carcinogenic hydrocarbons. Proc. Rudolf Virchow med. Soc. City N. Y. (Suppl.) 26, 394 (1968).
19. CUSHING, H.: Personal communication.
20. DRUCKREY, H., IVANKOVIC, S., PREUSSMANN, R., ZÜLCH, K. J., MENNEL, H. D.: Selective induction of malignant tumors of the nervous system by resorptive carcinogens. In: The Experimental Biology of Brain Tumors, p. 85. KIRSCH, W. M., PAOLETTI, E. G., and PAOLETTI, P. (eds.) Springfield, Ill.: Charles C. Thomas 1972.
21. KOESTNER, A., SWENBERG, J. A., WECHSLER, W.: Experimental tumors of the nervous system induced by resorptive N-nitrosourea compounds. Progr. exp. Tumor Res. 17, 9 (1972).
22. KOESTNER, A., SWENBERG, J. A., WECHSLER, W.:Transplacental production with ethylnitrosourea of neoplasms of the nervous system in Sprague-Dawley rats. Am. J. Pathol. 63, 37 (1971).
23. IKUTA, F., KUMANISHI, T.: Experimental virus-induced brain tumors. In: Progress in Neuropathology, Vol. 2. p. 253. ZIMMERMAN, H. M. (ed.). New York: Grune & Stratton 1973.
24. MUKAI, N., KOBAYASHI, S.: Primary brain and spinal cord tumors induced by human adenovirus type 12 in hamsters. J. Neuropathol. Exptl. Neurol. 32, 523 (1973).
25. ZIMMERMANN, H.M.: In explantion of certain glioma problems. J. M. Sinai Hosp. N. Y. 24, 1357 (1957).
26. ZIMMERMAN, H. M., ARNOLD, H.: Experimental brain tumors. I. Tumors produced with methylcholanthrene. Cancer Res. 1, 919 (1941).
27. ZIMMERMAN, H. M.: Experimental brain tumors. In: The Biology and Treatment of Intracranial Tumors, p. 49. FIELDS, W. S. and SHARKEY, P. C. (eds.). Springfield, Ill.: Charles C. Thomas 1962.

Refinement of the Avian Oncornavirus-Induced Primary Rat Brain Tumor Model for Therapeutic Screening

D. D. Bigner, D. J. Self, J. Frey, R. Ishizaki, A. J. Langlois, and J. A. Swenberg

INTRODUCTION

The relevance of animal models for human disease is a topic often hotly debated by experimentalists, especially when the animal model is to be used to evaluate unproven and potentially hazardous therapeutic innovations prior to initial clinical investigation. For decades general oncologists (1, 2) have argued for the use of primary, autochthonous tumors rather than transplanted tumors on the basis of 1) potential or proven differences between transplanted tumor and host histocompatibility antigens; 2) the possibility of tumor contamination by virus or mycoplasma during transplant; and 3) the selection of cells during serial transplantation possessing growth properties unlike those of primary tumors. Moreover, neuro-oncologists (3, 4, 5) have felt that most transplanted brain tumor models differed greatly from the histological type of the most serious human brain tumor problem, the glioblastoma multiforme - anaplastic astrocytoma tumor group, and that blood supply and blood-brain barrier alterations were markedly different in transplanted and primary brain tumors. Nevertheless, the majority of therapeutic trials in brain tumor animal models, both chemotherapeutic and immunotherapeutic, have been conducted with transplanted tumors (6 - 20). One of the commonly used murine transplanted tumors, an ependomyoblastoma, was induced over 30 years ago and is now expressing large amounts of infectious murine mammary tumor virus (21). Another widely used cell culture and transplanted tumor line, the C_6 rat astrocytoma, undergoes sarcomatous change on transplant (22, 23); moreover, this tumor was induced in random bred Wistar-Furth rats and although it will grow in some Fischer 344 and Wistar-Furth inbred strains, it contains histocompatibility differences sufficiently strong to elicit homograft rejection responses against the transplant thereby falsely enhancing any experimental therapeutic effects (24, 25, 26). Transplanted tumors probably have been used so widely in neuro-oncology because of the relative ease in maintaining and using such models, compared to the relatively long latency periods and scattered survival periods observed with chemically or radiation-induced tumors, and the general unavailability of oncogenic viruses suitable for reproducible experimental brain tumor induction.

The present study was undertaken to determine if the model of primary, autochthonous Avian Sarcoma Virus (ASV)-induced mammalian brain tumors could be refined for use in experimental chemotherapy and immunotherapy. It was found that large scale cell culture production and concentration of standard lots of infectious cell-free homogeneous subgroup C Bratislava-77 ASV (B-77-C-ASV) or cloned homogeneous subgroup D Schmidt-Ruppin (SR-D) ASV was feasible, and that intracerebral inoculation of such concentrated ASV into neonatal inbred Fischer 344 rats induced a 100% incidence of anaplastic astrocytomas. Moreover, nearly all inoculated animals died from brain tumors with highly reproducible death rates so that statistical evaluation of the effects of chemotherapy and immunotherapy on survival was possible.

MATERIALS AND METHODS

B77-ASV was obtained as a 10% chicken tumor extract from Dr. JOSEPH BEARD, who had passed the virus twice in line 15 chickens after obtaining it from Dr. ELENA HLAVAYOVA of Bratislava. A B-77-C standard strain (B-77-V) was obtained from Dr. P. K. VOGT and recloned three times on C/ABE chicken fibroblasts by the method of GRAF et al. (22). SR-D-ASV which had been rescued from a hamster tumor, cloned three times on C/AE chick fibroblasts, and shown to be homogeneous subgroup D virus by host range, interference, and neutralization testing was obtained from Dr. HEINZ BAUER (28). Virus was assayed by focus formation on group specific (GS) antigen negative C/E chick secondary or tertiary fibroblasts by the method of VOGT (29) and foci were enumerated by the method of SELF et al. (30). Virus particle counts were done by the method of SHARP and BEARD (31). A standard sample of B77-V was titrated with each set of unknowns to control for variation within or between assays (32). For production of B77-C an infecting stock virus was prepared by co-cultivating a hamster glioma induced with the Beard-Hlavayova B77-C with C/E chick fibroblasts. The infecting B77-C pool and its progeny were tested for sub-group C homogeneity by host range, interference and neutralization testing (33, 34, 35). Roller bottle production of B77-C or SR-D ASV was carried out by initiating infection of 1 x 10^6 C/E GS antigen negative chick fibroblasts on 100 mm dishes at a multiplicity of infection of 0.001 and transferring cultures at confluency at a density of 6 x 10^7 cells (usually about 30% transformed) to 670 mm roller bottles (36). Medium was changed daily until at least 75% transformation occurred after which the roller bottles were harvested twice daily and processed as described in results. Inbred Fischer 344 rats were obtained from Charles River Laboratories (CDF strain) and inoculated intracerebrally on the first postnatal day as previously described (37). Complete autopsies were performed and tumors and organs were examined by light and electron microscopy (38). Statistical evaluation of survival between litters was done by Wilcoxon rank sum analysis (39). Intravenous chemotherapy with 1,3 bis (2-chlorethyl)-1-nitrosurea (BCNU) and 2(2-chlorethyl)-3-cyclo-hexyl-1-nitrosurea (CCNU) was carried out as decribed by SWENBERG et al. (40). Immunosuppression with horse anti-rat thymocyte antiserum was carried out with the same sera and dose schedules described by DENLINGER et al. (41). Immunotherapy was given by administering 10^8 live Mycobacterium bovis (BCG) organisms intradermally, intracerebrally, or intradermally mixed with syngeneic B77-C-transformed rat sarcoma cells irradiated with 1,000 R.

RESULTS

Subgroup Homogeneity of B-77-C-ASV

The infecting pool of B-77-C rescued from a B-77-C induced hamster sarcoma by co-cultivation with chick fibroblasts, as well as virus produced from this stock was tested for subgroup C homogeneity by interference tests, host range, and neutralization properties. Interference with B-77-C focus formation was observed only with cells pre-infected with RAV-7 subgroup C leukosis virus, and not with leukosis viruses of subgroups A, B or D (Table 1). In host range testing the B-77-C plated with equal efficiency on C/E, C/AE, C/BE, and C/ABE cells, which excluded viruses of the other major subgroups (Table 2). There was complete neutralization of all three B-77-C preparations by subgroup C antisera but no reduction in titer with subgroup A, B, D, or E neutralizing antisera (Table 3).

Table 1. Interference test of B77-C-ASV on C/E GS antigen negative cells

Interfering rous-associated (leukemia) virus		Relative plating efficiency of challenge sarcoma virus[a]			
		RSV (RAV-1)	RSV (RAV-2)	B77	RSV (RAV-50)
Virus	Subgroup	A	B	C	D
None	--	1.00	1.00	1.00	1.00
RAV-1	A	<0.001	3.33	1.14	1.61
RAV-2	B	0.71	<0.01	0.61	<0.01
RAV-7	C	1.03	0.96	<0.01	1.03
RAV-50	D	0.78	0.04	0.77	0.05

[a] Ratio of titer on RAV-infected cells/uninfected cells.

Table 2. Relative plating efficiencies of B77-C-ASV on phenotypically different chicken cells

Avian sarcoma virus		Cells and titer (FFU/ml)			
Subgroup	Virus	C/E	C/AE	C/BE	C/ABE
A	RSV(RAV-1)	6.4×10^5	$< 1 \times 10^1$	5.2×10^5	$< 1 \times 10^1$
B	RSV(RAV-2)	1.4×10^5	1.9×10^5	$< 1 \times 10^1$	$< 1 \times 10^1$
C	B-77	8.6×10^4	8.1×10^4	8.2×10^4	8.1×10^4
D	Schmidt-Ruppin D	4.0×10^5	4.1×10^5	$< 1 \times 10^1$	$< 1 \times 10^1$
E	RSV_B(F +)[a]	$< 1 \times 10^1$	$< 1 \times 10^1$	$< 1 \times 10^1$	$< 1 \times 10^1$

[a] Titered 2.2×10^4 FFU/ml on quail cells.

Optimal Methods for Cell Culture Production and Concentration of ASV for Experimental Brain Tumor Induction

It has been shown by VOGT et al. (42) that dimethylsulfoxide (DMSO) enhanced virus replication of some ASV subgroups. No significant differences were observed in the amount of infectious B-77 produced in stationary monolayer culture with medium containing 1% or 2% DMSO (Fig.1). The only observed effect of DMSO in stationary culture was a slight reduction in growth rate of cells in 2% DMSO. Saturation cell densities

Table 3. Neutralization properties of B77-C-ASV. FFU/ml after incubation with subgroup specific antisera

Virus	Virus dilution	Antisera subgroup				
		A	B	C	D	E
B-77-C infecting pool	10^{0}	21	9	0	18	22
	10^{-1}	0	0	0	0	0
Concentrated B-77-C	10^{-5}	100	100	0	100	100
	10^{-4}	>500	>500	0	>500	>500
Concentrated B-77-C-Vogt	10^{-5}	120	100	0	100	100
	10^{-4}	>500	>500	0	>500	>500

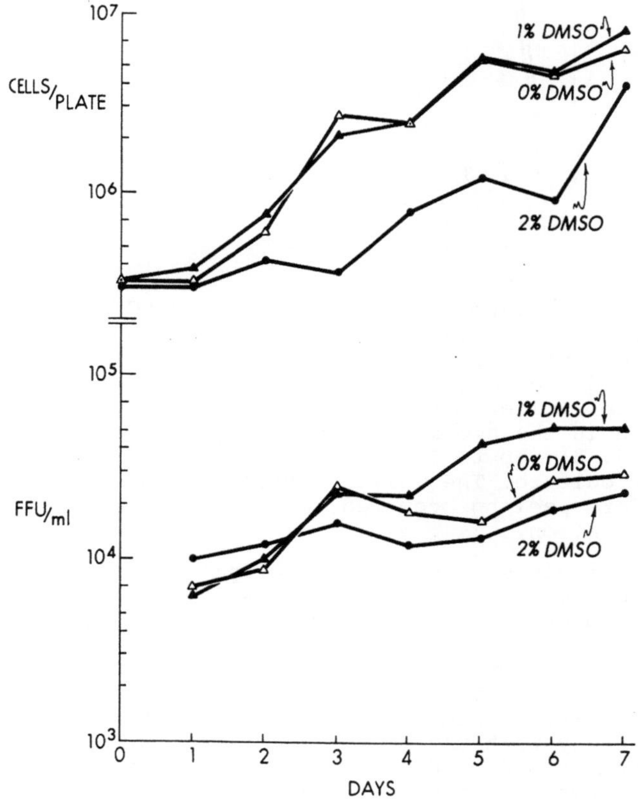

Fig. 1. Effect of 0, 1%, or 2% DMSO on cells and B77-C-ASV production in 60 mm plates

in medium with 0%, 1%, or 2% DMSO were similar. Nevertheless, in roller culture output of infectious virus was up to 10 times higher in 1% DMSO

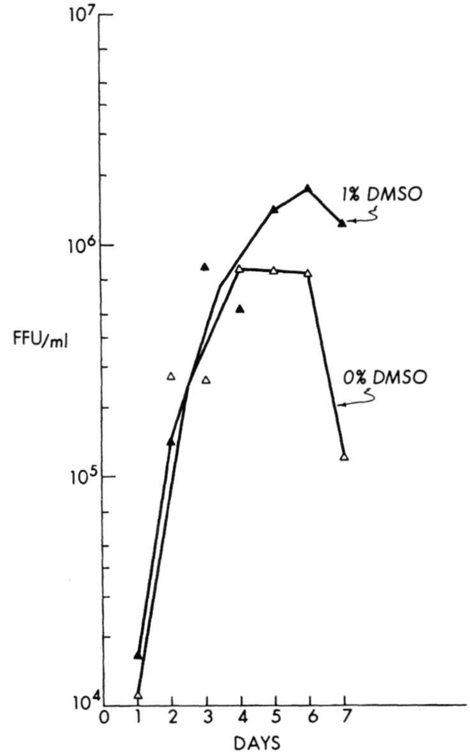

Fig. 2. Effect of DMSO on B77-C-ASV production in roller bottles

medium compared to medium without DMSO (Fig. 2). Moreover, cells in roller bottles without DMSO began peeling and deteriorating 6 to 7 days after initiation, whereas, roller bottle cultures with 1% DMSO have remained stable for over 3 weeks with sustained high output of infectious virus. Roller bottle cultures were superior to stationary plate cultures, both in the amount of physical particles and infectious virus produced (Fig. 3). Up to 15 times more virus per milliliter was produced in roller culture. The roller bottles could be maintained for weeks or months before peeling occurred, whereas plates usually peeled 7-10 days after initiation.

Polyethylene glycol (PEG) precipitation was evaluated as an initial step in concentration of infectious ASV from culture supernatants because of the potential of PEG precipitation for more rapid concentration of large volumes of culture fluids than could be obtained with ultracentrifugation (43). In agreement with similar studies (36) using a different subgroup C-ASV, Prague strain virus, it was found that only 25% of viral infectivity recovery was observed with PEG precipitation compared to 42% recovery with ultracentrifugation (Table 4). Therefore a concentration scheme based on differential centrifugation and filtration with a minimal number of freeze-thawing steps was developed (Fig. 4). With this concentration scheme, several pools of 1000-5000 one-hundred-percent brain tumor inducing doses (2-20 X 10^5 FFU/0.005 ml) of B-77-C and SR-D have been prepared. These pools have not lost activity during storage at -70° C for over 12 months. The concentrated virus consisted of typical C-type virus particles and subcellular particles of the transformed cells that were disrupted during the concentration process (Fig. 5).

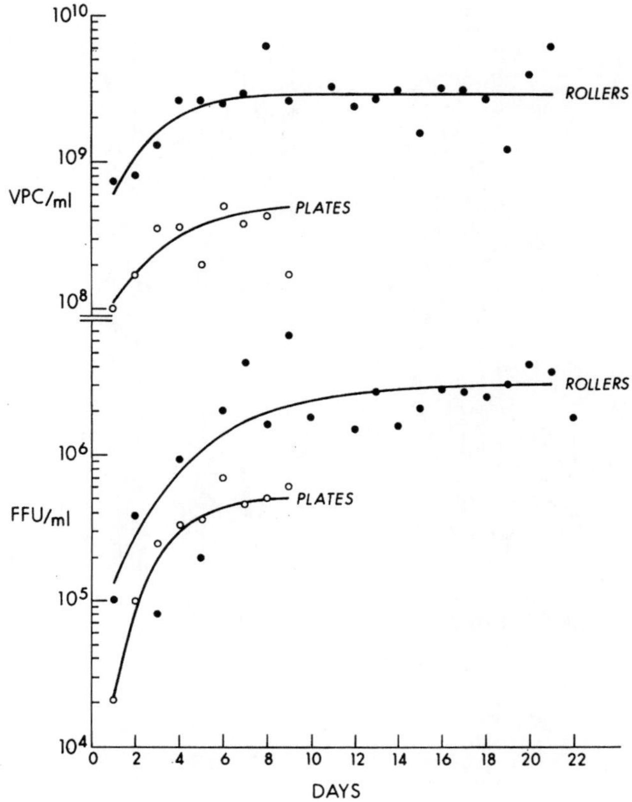

Fig. 3. Comparison of B77-C-ASV virus particle and focus forming unit production in roller bottles and plates

Table 4. Comparison of infectivity recovery after concentration of B77-C-ASV by centrifugation or PEG precipitation

Total FFU	Total FFU after concentration x 10^6 / Total FFU before concentration x 10^6	% Recovery
Centrifugation[a]	5.9 / 14.0	42%
PEG Precipitation[b]	4.8 / 19.2	25%

[a] 19,000 RPM (36,000 x G) in 19 head of L2-65B Beckman ultracentrifuge for 4 hrs.

[b] Stirred for 4 hr. at 4°C in solution of 10% PEG and 3% sodium chloride followed by 5500 RPM (5000 x G) spin in 845 head of IEC PR[b] centrifuge for 40 min.

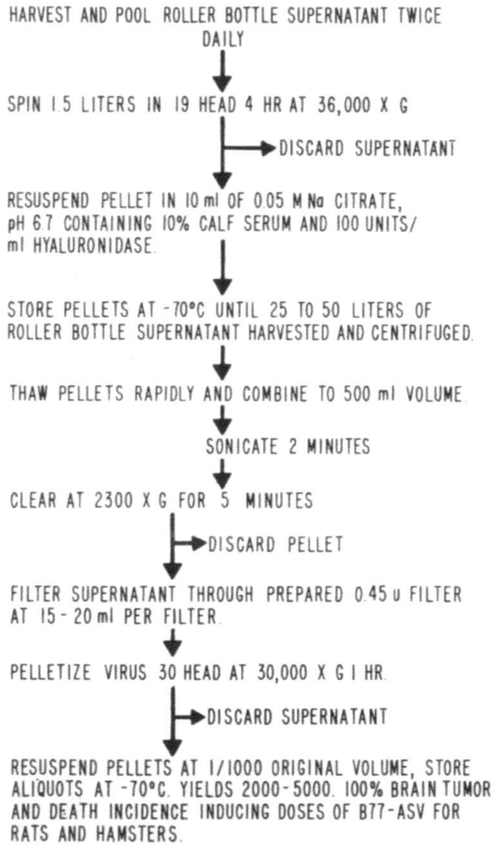

Fig. 4

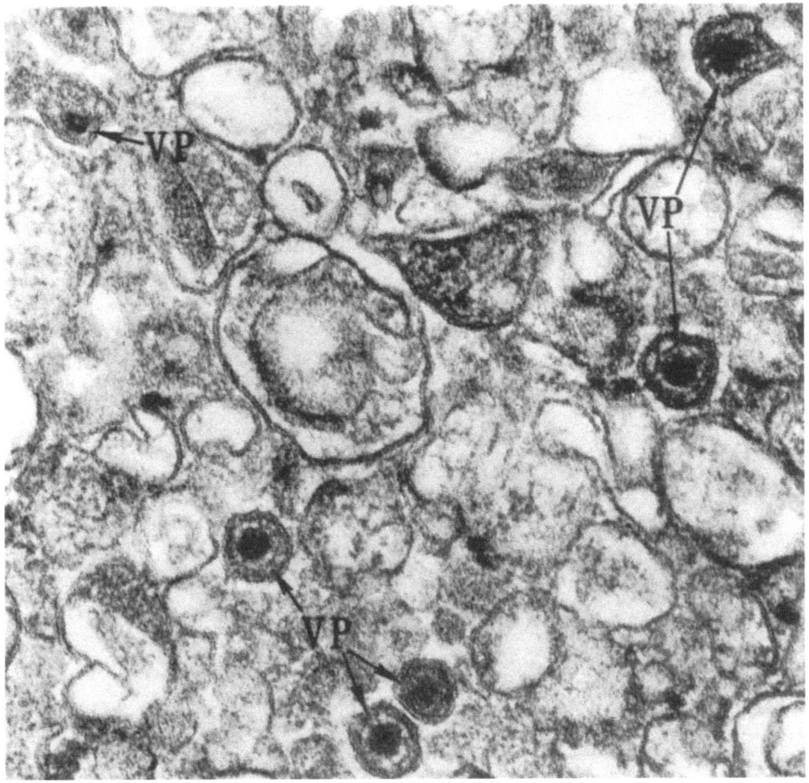

Fig. 5

Fig. 4. B77-C-ASV concentration scheme. Hyaluronidase was used in the buffer in which the original pellet was resuspended because of the large number of transformed cells that had been released into the roller bottle supernatants

Fig. 5. High resolution electron micrograph of concentrated B77-C-ASV pellet containing C-type virus particles and subcellular debris (X 100,000)

Brain Tumor Induction with Cell Culture Produced B77-C-ASV and SR-D-ASV

The majority of mammalian brain tumors induced with cell culture produced ASV in our laboratory has been with B-77-C. Hamsters, rats, and guinea pigs were all susceptible to intracerebral inoculation of B-77-C. Although a 100% incidence of brain tumors (astrocytomas, mixed gliomas, gliosarcomas, and sarcomas) was induced in hamsters and rats with B-77-C, only a 60% brain tumor incidence was observed in guinea pigs. Mean survival times ranged from 20 ± 7 days in hamsters to an intermediate value of 54 ± 11 days in Fischer 344 rats to no deaths in guinea pigs. Microscopic tumors were present in 6 of 10 inoculated guinea pigs when they were killed at 109 days after inoculation (Table 5).

Table 5. Mammalian brain tumor inducing range of B77-C-ASV

Strain and species	Dose FFU/inoculum[a] volume	Brain tumor incidence Gross	Microscopic	Death incidence	Mean survival time
PD_4 hamster	4.5×10^6/0.01 ml	100% (14/14)	100% (14/14)	100% (14/14)	20 ± 7 Days
Fischer 344 rat	4.8×10^6/0.01 ml	100% (54/54)	100% (54/54)	100% (54/54)	54 ± 11 Days
Hartley guinea pig	2.4×10^6/0.005 ml	0% (0/10)	60% (6/10)	0% (0/10)	>109 Days

[a] All inoculations were intracerebral at postnatal day 1 in rats and guinea pigs and day 3 in hamsters.

Table 6. Distribution of lesions induced in Fischer 344 rats after intracerebral inoculation of 1 day old neonates with B77-C-ASV

Brain tumors (gliomas)	Meningeal[a] tumors (sarcomas)	Spinal cord tumors	Hemorrhagic cysts
24/24	3/24	9/24	9/24
100%	12.5%	37.5%	37.5%

[a] 18/24 (75%) of animals had microscopic meningeal sarcomatosis.

Hamster brain tumors induced with B-77-C have been described in detail elsewhere (44) and included gliomas, distinctive large cell tumors of uncertain cell of origin, and sarcomas. Guinea pig B-77-C-induced brain tumors were all astrocytomas of the cortex. In the inbred Fischer 344 rat, the predominant brain tumors induced with B-77-C were infiltrative intracerebral astrocytomas or intraventricular astrocytic neoplasms

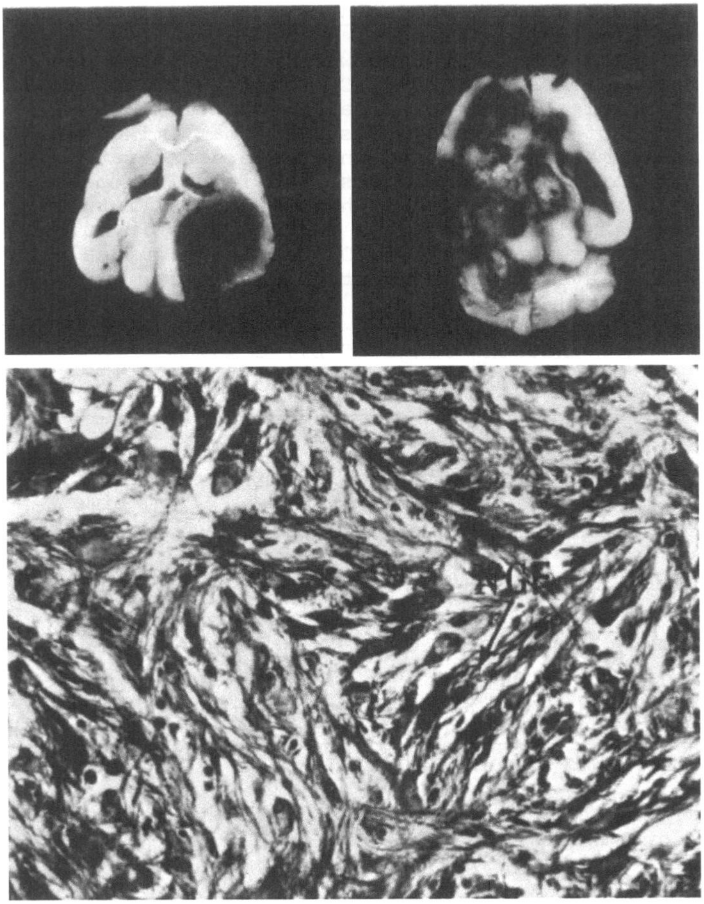

Fig. 6. Upper: gross photographs of B77-C-ASV induced Fischer 344 rat brain tumors. Lower: anaplastic astrocytoma with neuroglia fibrils (NGF), PTAH stain, X 400

(Fig. 6). Other lesions included intracerebral sarcomas and hemorrhagic cyst disease of rats which also has been described (45) following subcutaneous inoculation of ASV (Table 6). A 100% brain tumor incidence was also induced in rats, hamsters, and dogs with SR-D-ASV.

Preliminary morphologic analysis of the brain tumors induced in hamsters, rats, and dogs with cell culture produced SR-D-ASV has revealed no major differences from earlier studies in which tumor-extracted SR-ASV was used, or from B-77-C induced tumors (37, 46).

Reproducibility and Potential Use of the Model for Evaluating Therapeutic Regimens

The ideal brain tumor model for therapeutic investigations should consist of readily inducible tumors which are histologically similar to common human gliomas, possess the blood supply and growth characteristics of a primary brain tumor, and have a reasonably constant survival time so that statistical evaluation can be made between treatment and control groups. The ASV brain tumor model in either the Fischer 344 or Sprague-Dawley rat provides an animal system which fulfills these requirements. As described above, up to 100% of the rats inoculated as neonates developed astrocytomas. Survival time curves were remarkably constant when animals were randomized at the time of inoculation and reassigned to foster mothers. Even litters inoculated at different times from the same pool of B-77-C-ASV had no statistically significant litter to litter variation in survival time curves (Fig. 7).

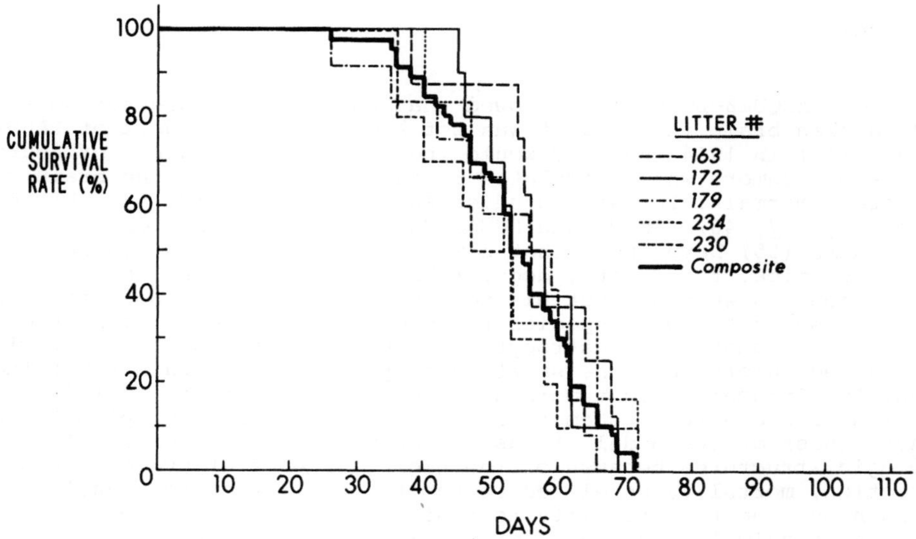

Fig. 7. Comparative mortality incidence of 5 litters of rats inoculated intracerebrally from a common pool of B77-C-ASV

A pilot chemotherapy study using the ASV model to evaluate the efficacy of BCNU or CCNU appeared quite promising (40). BCNU and CCNU were chosen for the initial chemotherapy studies since these drugs have already been shown to be effective in treating anaplastic astrocytomas and glioblastoma multiforme patients in clinical human trials. Survival times and histopathology of rats receiving intravenous dose levels of 5 or 10 mg/kg of each drug were compared to control rats receiving vehicle alone. All treated groups had increased median survival times compared to controls.

BCNU-treated rats survived 59-61% longer than controls (P 0.005), while CCNU only produced a 27-36% increase. Twenty-five percent of the BCNU-treated animals survived longer than the longest lived control rat, with one animal living more than four times the mean survival time of controls. This rat still died, however, of a brain tumor. Minimal toxicity was observed with the dose schedules employed.

On the other hand, immunosuppression as well as non-specific or specific immune stimulation had no statistically significant effect on survival of inbred Fischer 344 rats with ASV-induced brain tumors. Rats chronically immunosuppressed with ATS died at the same rate as rats receiving normal horse serum or saline alone. Rats given 10^8 organisms of Mycobacterium bovis (BCG), either intradermally, intracerebrally, or intradermally with irradiated syngeneic B-77-C-ASV rat sarcoma cells died at the same rate as untreated rats. Successful BCG infection had occurred as evidenced by specific elaboration of leukocyte migration inhibition factor by lymphocytes from treated animals in the presence of tuberculin purified protein derivative, as well as by the presence of BCG organisms in granulomas in inoculated rats (47). Nevertheless neither diminution or prolongation of survival occurred in the BCG treated rats, regardless of whether BCG was given 5 days before or 2 weeks after intracerebral virus inoculation.

DISCUSSION

Although VASQUEZ-LOPEZ (48) showed that ASV would induce sarcomas in the chicken brain in 1936, it was not known until the work of RABOTTI et al. (49) in 1964 that ASV would induce a high incidence of neuroectodermal tumors when inoculated intracerebrally into the brain of neonatal mammals. In later studies RABOTTI et al. (50, 51), BIGNER et al. (3, 37, 46), IKUTA and KUMANISHI (52, 53), JAENISCH et al. (54) and others (55) have shown that a wide range of mammals, including hamsters, mice, rats, cats, dogs, rabbits, guinea pigs, marmosets, and rhesus monkeys were susceptible to the neuro-oncogenic potential of ASV.Because of the wide species range of susceptible animals, the high incidence of anaplastic astrocytomas, the lack of production of virus in induced tumors, and the short latency period for brain tumor induction, ASV-induced brain tumors have been an attractive model for experimental brain tumor chemotherapy or immunotherapy. Despite the attractiveness of the model it has not been widely used except for morphologic structures because of the high incidence of hydrocephalus, infection, mortality unrelated to tumor, the lack of reproducible death incidences, and the inability to achieve a predictable 100% tumor incidence in animals inoculated with chicken tumor extracted ASV. BIGNER et al. (3) overcame the problems of hydrocephalus and infection and induced a 100% brain tumor incidence by filtering and concentrating tumor-extracted ASV before inoculation. Nevertheless, it was not possible even with the improved methods of preparing tumor extracted ASV to prepare large pools of ASV with uniform titer and stability on storage. Only a few brain tumor-inducing doses could be prepared in one concentration pool from tumor-extracted ASV. When several individually prepared batches of concentrated tumor-extracted ASV were thawed and combined to make one large pool, a marked drop in virus titer and brain tumor-inducing potential was observed, presumably because of inactivation of infectivity due to multiple freeze-thawings the virus had undergone during the concentration procedures (56).

It was postulated that ASV could be produced in cell culture in sufficient quantity and concentrated with a minimal number of freeze thawings to prepare large standard pools of ASV for experimental brain tumor induction. Moreover, cell culture produced virus offered the additional advantage of providing virus homogeneous with respect to subgroup which may be necessary in characterizing tumor cell-surface membrane antigens.

For initial studies B77-C was chosen because of its relative lack of cytopathic effect in cell culture. The optimal growth condition for B-77-C production was found to be roller bottle growth of transformed cells in medium containing 1% DMSO. Such conditions allowed harvest of many liters of high-titered culture supernatants, since cultures could be maintained and harvested twice daily for many weeks. Once satisfactory conditions of growth for the non-cytopathic B-77-C-ASV were achieved, it was possible to apply successfully similar methods to roller bottle growth of the more cytopathic SR-D-ASV. PEG precipitation, although quite effective in precipitating some other RNA tumor viruses such as avian myeloblastosis virus (57), was less effective than ultracentrifugation in achieving concentration of infectious B-77-C. Therefore, a concentration scheme was developed based on differential centrifugation and filtration with only 2 freeze-thawings of the virus suspensions before intracerebral virus inoculation.

With highly concentrated uniform virus pools that were stable on storage and suspended in 0.05 M sodium citrate buffer in which inoculation could be carried out without further virus processing, it was possible to induce 100% brain tumor and death incidence with B-77-C and SR-D-ASV in hamsters and inbred rats. If animals were randomized at birth and reassigned to foster mothers, uniform, reproducible death incidences were observed. This reproducibility suggested application of the model to therapeutic evaluation. A pilot study demonstrated that BCNU, and to a lesser extent CCNU treatment of the autochthonus viral-induced brain tumors resulted in a 27-61% increase in survival. These results closely paralleled reports of CCNU treatment of human gliomas (58) and the Brain Tumor Study Group's prospective randomized trials of BCNU (59). This model appears highly appropriate for future screening of therapeutic regimens since large numbers of animals can be rapidly inoculated, randomized and treated.

In contrast to previous studies using BCNU and CCNU against transplantable "gliomas", no cures were obtained, although 20 percent of the animals survived 2 to 4 times as long as the median survival of untreated animals. It is important to note, however, that these drugs rarely produce cures in man. The system may be more predictive of drug-induced toxicity than transplantable brain tumor models since the ASV-inoculated animals survived long enough to develop the delayed forms of toxicity associated with drugs like the nitrosoureas. The model offers additional advantages including the use of tumors possessing a morphology similar to one of the most common human brain tumors, the anaplastic astrocytoma, as well as the blood supply and growth characteristics of a primary brain tumor.

Similar to some chemically induced brain tumor model systems, immunosuppression neither shortened nor prolonged survival rates with ASV-induced rat brain tumors (41). Likewise, immunotherapy had no effect on survival. Such data underline the necessity for proceeding carefully with human brain tumor immunotherapy since immunotherapy may not only be ineffective but potentially harmful.

ASV-induced brain tumors in rats have been shown to be a primary brain tumor animal model of a common anaplastic human brain tumor type which may serve as a useful screen for chemotherapeutic or other novel brain tumor treatment methods before human clinical trials.

Acknowledgements

Dr. THOMAS GRAF performed the virus neutralization studies, and Dr. ROBERT DENLINGER aided in preparing and evaluating the horse antithymocyte antiserum. We thank KATHY PETTIS, PHILIP HARBACH, and THOMAS HALL for assistance.

REFERENCES

1. OBERLING, C.: The Riddle of Cancer, p. 55. New Haven: Yale University Press 1952.
2. DAY, E. D.: Animal Cancer, the primary tool: Guest editorial. Cancer Res. 21, 581 (1961).
3. BIGNER, D. D., KVEDAR, J. P., SHAFFER, T. C., VICK, N. A., ENGEL, W. K., DAY, E. D.: Factors influencing the cell type of brain tumors induced in dogs by Schmidt-Ruppin Rous sarcoma virus. J. Neuropathol. Exp. Neurol. 31, 583 (1972).
4. WECHSLER, W., KLEIHUES, P., MATSUMOTO, S., ZUELCH, K. J., IVANKOVIC, S., PREUSSMAN, R., DRUCKREY, H.: Pathology of experimental neurogenic tumors chemically induced during prenatal and postnatal life. Ann. N. Y. Acad. Sci. 159, 360 (1969).
5. SWENBERG, JAMES A., KOESTNER, A., WECHSLER, W.: The induction of tumors of the nervous system with intravenous methylnitrosurea. Lab. Invest. 26, 74 (1972).
6. CHIRIGOS, M. A., HUMPHREYS, S. R., GOLDIN, A.: Duration of effective levels of three antitumor drugs in mice with leukemia L1210 implanted intracerebrally and subcutaneously. Cancer Chemotherapy Rept. 49, 15 (1965).
7. KOTSILIMBAS, D. G., KARPF, R., MEREDITH, S., SCHEINBERG, L. C.: Evaluation of parenteral 5-FU on experimental brain tumors. Neurology 16, 916 (1966).
8. KOTSILIMBAS, D. G., MEYER, L., BERSON, M., TAYLOR, J. M., SCHEINBERG, L. C.: Corticosteroid effect on intracerebral melanomata and associated cerebral edema: Some unexpected findings. Neurology 17, 223 (1967).
9. ROSSO, R., DONELLI, M. G., REYEERS-DEGLI INNOCENTI, I., GARATTINI, S.: Chemotherapy of tumors transplanted intracerebrally. European J. Cancer 3, 125 (1967).
10. KENNEDY, B. J., YARBRO, J. W., KICKERTZ, V., SANDBERG-WOLLHEIM, M.: Effect of mithramycin on a mouse glioma. Cancer Res. 28, 91 (1968).
11. SHAPIRO, W. R., AUSMAN, J. I.: The chemotherapy of brain tumors: a clinical and experimental review. In: F. PLUM (ed.), Recent Advances in Neurology, pp. 149-235. Philadelphia: F. A. Davis Co. 1969.
12. WRIGHT, R. L., SHAUMBA, B., KELLER, J.: The effect of glucocorticosteroids on growth and metabolism of experimental glial tumors. J. Neurosurg. 30, 140 (1969).
13. AUSMAN, J. I., SHAPIRO, W. R., RALL, D. P.: Studies on the chemotherapy of experimental brain tumors: Development of an experimental model. Cancer Res. 30, 2394 (1970).
14. LEVIN, V. A., SHAPIRO, W. R., CLANCY, T. P., OLIVERIO, V. T.: The uptake, distribution, and antitumor activity of 1-(2-chloroethyl)-3-cyclohexyl-1-nitrosourea in the murine glioma. Cancer Res. 30, 2451 (1970).
15. SHAPIRO, W. R., AUSMAN, J. I., RALL, D. P.: Studies on the chemotherapy of experimental brain tumors: Evaluation of 1,3-Bis(2-chloroethyl)-1-nitrosourea, cyclophosphamide, mithramycin, and methotrexate. Cancer Res. 30, 2401 (1970).
16. GURCAY, O., WILSON, C. B., BARKER, M., ELIASON, J.: Cortico-steroid effect on transplantable rat glioma. Arch. Neurol. 24, 266 (1971).
17. SHAPIRO, W. R.: Studies on the chemotherapy of experimental brain tumors: Evaluation of 1(2-chloroethyl)-3-cyclohexyl-1-nitrosourea, vincristine, and 5-fluorouracil. J. Natl. Cancer Inst. 46, 359 (1971).
18. SHAPIRO, W. R.: Effects of 1-(2-chloroethyl)-3-cyclohexyl-1-nitrosourea (CCNU, NSC 79037) in sesame oil intramuscularly on experimental brain tumors. Cancer Chemother. Rep. 56, 457 (1972).

19. SHAPIRO, W. R.: The chemotherapy of intracerebral vs. subcutaneous murine gliomas. Arch. Neurol. 30, 222 (1974).
20. SHAPIRO, W. R., POSNER, J. B.: Corticosteroid hormones: Effects in an experimental brain tumor. Arch Neurol. 30, 217 (1974).
21. BARKER, R. C., AMES, R. P.: Mammary tumor virus in experimental ependymoblastoma. Progr. in Neuropathology II, 335 (1973).
22. BARKER, M., HOSHINO, T., GURCAY, O., WILSON, C. B., NIELSEN, S. L., DOWNIE, R., ELIASON, J.: Development of an animal brain tumor model and its response to therapy with 1, 3-Bis (2-chloroethyl)-1-nitrosurea. Cancer Res. 33, 976 (1973).
23. DAY, E. D., BIGNER, D. D.: Specificity, cross-reactivity, and affinity of ^{125}I-labeled antiglioma antibodies for monolayers of cultured glioma cells. Cancer Res. 33, 2362 (1973).
24. MESSER, J., SWEET, W.: Personal communication.
25. BIGNER, D. D.: Unpublished results.
26. WECHSLER, W.: Personal communication.
27. GRAF, T., BAUER, H., GELDERBLOM, H., BOLOGNESI, D.: Studies on the reproduction and cell-converting abilities of avian sarcoma viruses. Virology 43, 427 (1971).
28. BAUER, H., GRAF, T.: Evidence for the possible existence of two envelope antigenic determinants and corresponding cell receptors for avian tumor viruses. Virology 37, 157 (1969).
29. VOGT, P. K.: Focus assay of Rous sarcoma virus. In Fundamental Techniques in Virology, K. HABEL and N. P. SALZMAN (eds.), pp. 198-204. New York: Academic Press 1969.
30. SELF, D. J., PETTIS, K. F., BIGNER, D. D.: Improved techniques for staining, evaluating, and enumerating focus formation in viral infectivity assays. Appl. Microbiol. (1974).
31. SHARP, D. G., BEARD, J. W.: Counts of virus particles by sedimentation on agar and electron microscopy. Proc. Soc. Exptl. Biol. Med. 81, 75 (1952).
32. BRYAN, W. R.: Biological studies on the Rous sarcoma virus. II. Review of sources of experimental variation and of methods for their control. J. Natl. Cancer Inst. 16, 287 (1955).
33. VOGT, P. K., ISHIZAKI, R.: Reciprocal patterns of genetic resistance to avian tumor viruses in 2 lines of chickens. Virology 26, 664 (1965).
34. VOGT, P. K., ISHIZAKI, R.: Patterns of viral interference of the avian leukosis and sarcoma complex. Virology 30, 368 (1966).
35. ISHIZAKI, R., VOGT, P. K.: Immunological relationships among envelope antigens of avian tumor viruses. Virology 30, 375 (1966).
36. SMITH, R. E., BERNSTEIN, E.H.: Production and purification of large amounts of Rous sarcoma virus. Appl. Microbiol. 25, 346 (1973).
37. WILFONG, R. F., BIGNER, D. D., SELF, D. J., WECHSLER, W.: Brain tumor types induced by the Schmidt-Ruppin strain of Rous sarcoma virus in inbred Fischer rats. Acta Neuropath. (Berlin) 25, 196 (1973).
38. VICK, N. A., BIGNER, D. D., KVEDAR, J. P.: The fine structure of canine gliomas and intracranial sarcomas induced by the Schmidt-Ruppin strain of the Rous sarcoma virus. J. Neuropath. Exptl. Neurol. 30, 354 (1972).
39. GEHAN, E. A.: A generalized Wilcoxon test for comparing arbitrarily singly censored samples. Biometrika 52, 203 (1965).
40. SWENBERG, J. A., BIGNER, D., HALL, T. L.: Experimental chemotherapy of viral-induced brain tumors. Proc. Amer. Assn. Canc. Res. 15, 42 (1974).
41. DENLINGER, R. H., SWENBERG, J. A., KOESTNER, A., WECHSLER, W.: Differential effect of immunosuppression on the induction of nervous system and bladder tumors by N-methyl-N.nitrosurea. J. Natl. Cancer Inst. 50, 87 (1973).

42. VOGT, P. K., TOYOSHIMA, K., YOSHI, S.: Factors promoting avian tumor virus infections. Proc. 2nd Int. Symp. Tumor Viruses, p. 229-238. Paris: Editions du centre national de la recherche scientifique 1970.
43. CLARK, M. F., LESTER, R. M.: The application of polyethylene-glycol solubility - concentration gradients in plant virus research. Virology 43, 338 (1971).
44. BURGER, P. C., BIGNER, D. D., SELF, D. J.: Morphologic observations of brain tumors in PD4 hamsters induced by 4 strains of avian sarcoma virus. Acta Neuropath. (Berlin) 26, 1 (1973).
45. ZILBER, L. A., KRIUKOVA, I. N.: Haemorrhagic disease of rats caused by Rous sarcoma virus. Vopr. Virusol. 4, 239 (1957).
46. BIGNER, D. D., ODOM, G. T., MAHALEY, M. S., Jr., DAY, E. D.: Brain tumors induced in dogs by the Schmidt-Ruppin strain of Rous sarcoma virus. J. Neuropathol. Expt. Neurol. 28, 648 (1969).
47. ADAMS, D. O., ROBINSON, S. C., BIGNER, D. D.: Unpublished results.
48. VAZQUEZ-LOPEZ, E.: On the growth of Rous sarcoma inoculated into the brain. Am. J. Cancer 26, 29 (1936).
49. RABOTTI, G. F., RAINE, W. A.: Brain tumors induced in hamsters inoculated intracerebrally at birth with Rous sarcoma virus. Nature 204, 898 (1964).
50. GROVE, A. S., Jr., DICHIRO, G., RABOTTI, G. F.: Experimental brain tumors with a report of those induced in dogs by Rous sarcoma virus. J. Neurosurg. 26, 465 (1967).
51. RABOTTI, G. F.: The Experimental Biology of Brain Tumors, Chapter 4. pp. 148-180. Springfield, Ill.: Charles C. Thomas 1972.
52. KUMANISHI, T.: Brain tumors induced with Rous sarcoma virus, Schmidt-Ruppin strain. I. Induction of brain tumors in adult mice with Rous chicken sarcoma cells. Japan J. Exptl. Med. 37, 461 (1967).
53. IKUTA, F., KUMANISHI, T.: Experimental virus-induced brain tumors. Progr. in Neuropath. II, 253 (1973).
54. JAENISCH, W., HORN, K. H., SCHOLTZE, P., SCHREIBER, D.: Attempt at induction of brain tumors in newborn Rhesus monkeys by intracerebral injection of Rous sarcoma virus (Ger.) Exptl. Pat. 2, 226 (1968).
55. DEINHARDT, F.: Rous sarcoma virus-induced neoplasms in new world non-human primates . Perspectives. Virol. 5, 183 (1967).
56. BIGNER, D. D.: Unpublished results.
57. LANGLOIS, A. J., Bolognesi, D. P.: Personal communication.
58. FEWER, D., WILSON, C. B., BOLDREY, E. B., ENOT. J. K.: Phase II study of 1-(2-chlorethyl)-3-cyclohexyl-1-nitrosurea (CCNU) in the treatment of brain tumors. Cancer Chemother. Reports 56, 421 (1972).
59. WALKER, M. D., GEHAN, E. A.: An evaluation of 1,3-bis (2-chlorethyl) -2-nitrosurea (BNCU) and irradiation alone and in combination for the treatment of malignant glioma. Proc. Am. Assoc. Cancer Res. 13, 67 (1972).

Studies on the Differentation of Human and Experimental Gliomas in Organ Culture Systems*

L. J. RUBINSTEIN and M. M. HERMAN

INTRODUCTION

Tissue culture techniques have been used for many years for the study of tumors of the nervous system (1-3). They have contributed a great deal of information bearing mostly on problems of cell morphology and, to some extent, cytogenesis. This is because, by their very nature, the technical advantages and limitations of these methods effectively restrict them to a largely cytologic approach in the study of tumors. Since the explants, covered by the fluid nutrient medium, are generally grown on a flat impervious glass surface, either with or without an adherent substrate such as plasma clot or reconstituted collagen, the tissue culturist necessarily has his attention focused on the proliferation and movement of individual cells that have migrated from the primary explant in a relatively two-dimensional direction. The great advantage is that the method of observation permits the optical study of whole cells, with an emphasis on their dynamic properties and their modulations in form with the passage of time and in response to regularly controlled changes in their environment.

The objective of organotypic, or organ culture, methods is different. Here the purpose is to reproduce as faithfully as possible the three-dimensional patterns of histologic organization and differentiation found in vivo. As originally designed for the study of embryonal organ rudiments and immature neonatal tissues, and later extended to normal mature or neoplastic tissues, organ culture techniques are based on the principle of culturing explants in a moist gas or air phase on the surface of a relatively large volume of stationary nutrient medium. A wide medium-gas interface is thus obtained, which results in an increased rate of oxygen diffusion into the culture. One of the consequences is that cellular migration to form a peripheral monolayer (which is characteristic of the more usual tissue culture systems) is to a large extent inhibited. Another consequence is that the proliferation and outgrowth of the less differentiated cell elements, in particular fibroblasts, tend also to be suppressed. The explant, after being maintained as a solid organized structure in vitro for a variable length of time, can then be fixed, embedded and processed in the conventional manner for light or electron microscopy. Other morphologic techniques, such as histochemistry and cytochemistry, immunofluorescence and autoradiography

---

*Supported by Research Grant CA-11689 from the National Cancer Institute, U.S.P.H.S. The initial stages of this work were supported also by grants from the American Cancer Society (Special Grant 504 from the California Division and Institutional Grant IN 32J) and from the Damon Runyon Memorial Fund for Cancer Research (DRG 1092).

may also be employed. Such an in vitro system can therefore be a useful tool for the study of morphogenesis and organogenesis. In the case of neoplasms, the method is more relevant to the study of histologic patterns of organization found in the solid state than the more traditional tissue or cell culture techniques. It could provide a greater degree of validity in the correlations to be made between in vivo and in vitro events connected with the morphologic expressions of differentiation and anaplasia. Finally, as suggested by LUMSDEN (1), it might constitute a long-term model with more reliable morphologic criteria for chemotherapeutic and radiotherapeutic screening than monolayer cell culture techniques.

METHODS

Several variants of the organ culture method have been described. Essentially they fall into two groups. The first method consists in supporting the explants on a fluid medium by means of a supporting flat and relatively rigid platform, or raft, which is frequently perforated to allow for better gas and nutrient interchange. Because a plasma clot or collagen substrate is lacking, the growth of cells by contact guidance (or "thigmotropism") to form a peripheral monolayer is inhibited. The explants tend to "epithelialize" themselves along their perimeter, where the outermost layer of cells becomes flattened. The second technique is based on the use of porous three-dimensional matrices, in which the nutrient medium circulates and is retained by capillary action in the lacunae of the matrix. This favors cell growth and migration in three directions, in particular within the interstices of the matrix. The advantage of the system is that, in the case of neoplasms, the property of invasiveness by the explant may be studied in vitro and compared to the invasive characteristics displayed by the tumor in vivo.

In the past four years, we have investigated a number of human and experimental nervous system tumors by means of such organotypic culture techniques. As a model for the raft method, we have used a Millipore filter method similar to that used by cell biologists for the study of inductive morphogenetic factors that control tissue interactions in developing embryonal rudiments. A Millipore filter (25 microns in thickness and 0.45 micron in pore size) is cemented over the central hole drilled in a small rectangular plastic sheet that fits across the central well of a Falcon organ culture dish. For the three-dimensional matrix technique, we have used pieces of gelatin sponge foam or human fibrin foam supported on stainless steel grids. In all experiments, sister cultures from the same tumors were grown on collagen-coated coverslips placed at the bottom of small Petri dishes, so as to monitor the state of the living cultures by phase contrast microscopy and compare the histologic features of the organ cultures with those seen with the more conventional tissue culture method. Cultures are maintained as an open system in a CO_2 incubator that provides an atmosphere of 3% to 4% CO_2, 96% of air, and a relative humidity of 97%. At regular (usually weekly) intervals the cultures are fixed and processed for light and electron microscopy. Details of the culture techniques, conditions of incubation, and monitoring and tissue processing methods have been reported (4, 5).

MATERIALS

We have to date attempted to culture 38 nervous system tumors obtained from patients at the time of craniotomy or laminectomy, and 14 experimental neoplasms which developed in the progeny of rats given a single intravenous injection of ethylnitrosourea (ENU) on the 15th or 16th day of pregnancy. In the case of human tumors, we have successfully maintained 28 of these neoplasms by organ culture. Our results largely reflect our interest in selecting for culture tumors of the glioma group. Table 1 lists the number and types of human and experimental gliomas in which culture was attempted and those in which it was successful. In addition to the 21 successfully cultured human gliomas listed, the following non-glial intracranial tumors were also successfully cultured: three pituitary adenomas, one acoustic Schwannoma (6), one cerebellar capillary hemangioblastoma (7), one chordoma, and one metastatic carcinoma. In the case of ENU-induced rat nervous system tumors, 8 of the successfully cultured 11 tumors were malignant Schwannomas originating from the cranial or spinal nerve roots; these will be the subject of detailed reports elsewhere (6, 8, 9). In this report, we shall review our chief findings in human and experimental gliomas.

Table 1. Attempts at maintenance of gliomas in organ culture systems

A. Human gliomas	Attempted	Successful
Glioblastomas	11[a]	10
Cerebral astrocytomas	9	4[b]
Cerebellar astrocytomas	3	3
Mixed astrocytoma and oligodendroglioma	1	1
Medulloblastoma	1	1
Oligodendroglioma	1	0
Ependymomas[c]	2	2
Ganglioglioma	1	0
Total	29	21
B. ENU-induced rat gliomas	6	3[d]

[a] One failure due to delay in explantation.
[b] All the successful cultures were obtained during the later period of this project.
[c] One myxopapillary ependymoma of the filum terminale, and one cellular 4th ventricle ependymoma.
[d] One spinal and two cerebral astrocytomas.

RESULTS

The histologic types of human and experimental gliomas successfully maintained in organ culture are listed in Tables 1 and 2. As expected, the glioblastomas were most easily maintained, successful culture being obtained in almost all cases. They were also maintained the longest, many cultures remaining viable for more than two months, and, in the case of one tumor (T-10), for over four months. In the case of less malignant gliomas, viability was usually maintained on sponge foam

Table 2. Maximal maintenance period of gliomas in tissue and organ culture systems

A. Human	On coverslips	On sponge foam	On Millipore filter
Glioblastomas (4, 5)	75 days	137 d.	122 d.
Cerebral astrocytomas	35 d.	42 d.	35 d.
Cerebellar astrocytomas (5)	37 d.	43 d.	29 d.
Medulloblastoma (11)	17 d.	22 d.	32 d.
Ependymoma			
Myxopapillary (10)	48 d.	69 d.	34 d.
4th ventricle (13)	51 d.	86 d.	46 d.
Mixed astrocytoma and oligodendroglioma	88 d.	67 d.	31 d.
B. ENU-induced rat			
Astrocytomas (12)	27 d.	55 d.	19 d.

matrices from one to two months, and often up to approximately one month on Millipore filters. The growth and maintenance of explants on Millipore filters was rather less successful than on the three-dimensional matrices, many cultures on Millipore filters undergoing either total or subtotal necrosis or sclerosis, with only a residual rim of viable tumor cells persisting along the periphery of the explant. Well-differentiated gliomas that were successfully cultured include 4 out of 9 cerebral astrocytomas, 3 out of 3 attempted cerebellar astrocytomas, and 2 ependymomas, one of which was a myxopapillary ependymoma originating in the filum terminale and the other a typical cellular ependymoma arising in the fourth ventricle. Failure to maintain an oligodendroglioma in organ culture should be noted, despite the fact that we were able to grow the same tumor on collagen-coated coverslips. Likewise, we have so far been unable to grow an experimental ENU-induced rat oligodendroglioma, although we have been successful in maintaining 3 experimental gliomas that were largely astrocytic.

The growth characteristics and in vitro evolution of these tumors in organ culture systems have been detailed in various publications from this laboratory (4, 5, 10-13). We here present an overall survey of the results of our studies and shall attempt to draw a few general conclusions of biological interest that can be derived from these collective observations.

As shown in Figs. 1 to 4 and as illustrated elsewhere (4, 5, 10), organotypic culture methods can successfully be used for the growth and maintenance of human and experimental gliomas in vitro, and the patterns of histologic organization and differentiation seen in vivo are remarkably well retained in these systems. In the case of gelfoam or human fibrin foam matrices, the technique provides an excellent scaffold that permits the explant to demonstrate its invasive properties, which are particularly striking in the more malignant gliomas, such as the glioblastomas. Invasion of the matrix is easily recognized by the presence of small particles of foam that have been segregated among the tumor cells. Invasiveness is also exhibited, although to a lesser degree, by less aggressively growing tumors of the glioma group, such as cerebral and cerebellar astrocytomas, and ependymomas. It was displayed only to a slight and rare extent in the case of a cerebellar capillary hemangioblastoma, and not at all in the case of a human

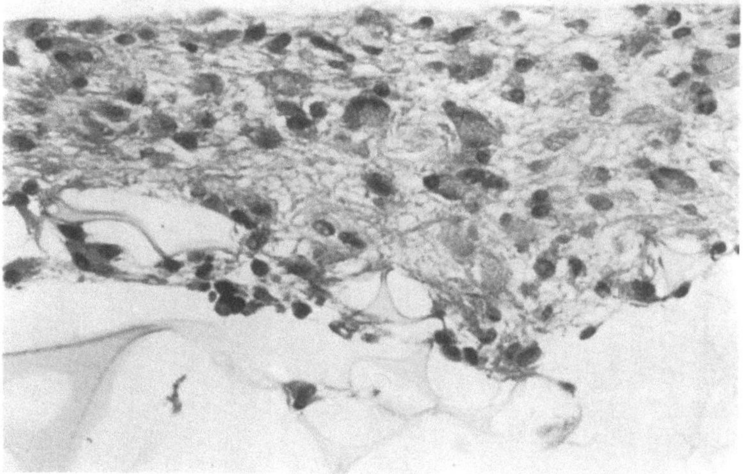

Fig. 1. Cerebellar astrocytoma cultured on sponge foam, 29 days in vitro. Note invasion of sponge foam, numerous glial fibers and well differentiated astrocytes. H & E x 300

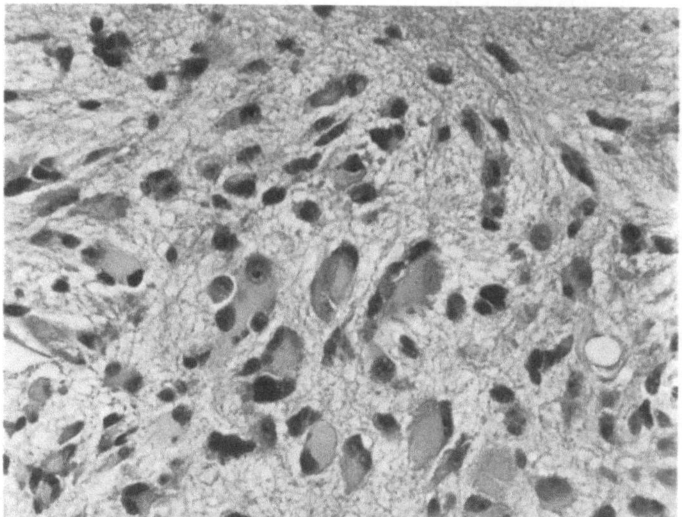

Fig. 2. Cerebral astrocytoma cultured on sponge foam, 21 days in vitro. Note prominent gemistocytic astrocytes and glial fibers. H & E x 300

acoustic Schwannoma. This lack of invasiveness of the acoustic Schwannoma contrasted markedly with the behavior in vitro of several malignant experimental ENU-induced Schwannomas (6).

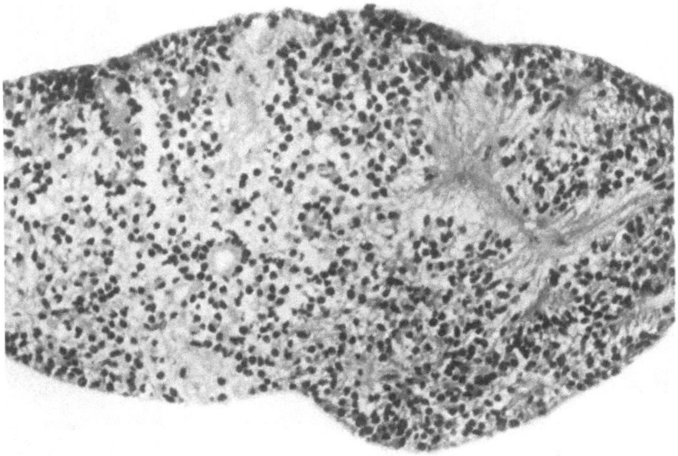

Fig. 3. Ependymoma from fourth ventricle cultured on Millipore filter, 30 days in vitro. Note well-formed ependymal rosettes and area rich in glial fibrils. H & E x 135

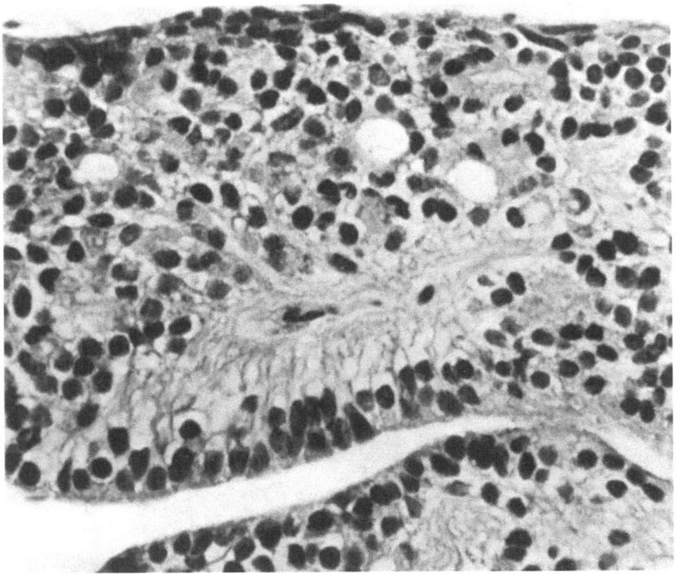

Fig. 4. Same tumor as Fig. 3, cultured on sponge foam, 30 days in vitro. Note perivascular pseudorosette (center), flattened layer of cells on medium-air interface (top), and well differentiated ependymal cells along lower edge of the explant (bottom). H & E x 375

As it has long been known, the distinction in these in vitro systems between actual growth and simple maintenance of viable explants is not always easy, but an increase in volume of the explants is usually quite evident within the first few weeks of culture. In the case of glioblastomas, the explants usually attained their maximal size within 3 to 4 weeks and maintained that size for up to 6 to 7 weeks, and in one case up to 122 days. After 6 to 7 weeks, most cultures begin to undergo a gradual decrease in volume and become ultimately reduced either to a thin cell layer on the surface of the matrix or to a relatively acellular sclerotic mass. In favor of active growth as opposed to simple maintenance are the presence of mitotic figures and the invasion of the three-dimensional matrix. Mitotic figures were usually most frequently seen in glioblastomas throughout the various stages of culture, and up to 122 days in vitro in one case. Kinetic studies designed to measure the growth fraction and the cell cycle time of the tumors maintained in these systems should provide important relevant information.

A number of general conclusions on the biological behavior of gliomas in organotypic culture systems can be drawn that have a bearing on the subject of tumor cell differentiation, histologic organization, and the in vitro reaction of the vascular and other mesenchymal stromatous elements.

1. Gliofibrillogenesis and Synthesis of Glial Fibrillary Acidic Protein

The first conclusion to be derived is that these in vitro systems maintain and accentuate the differentiating features of the explanted neoplasms. This is well demonstrated in tumors of the glioma group (Figs. 1 to 4). The tendency of organ culture to favor differentiation occurs irrespectively of the anaplastic character of the original neoplasm. Thus, biologically very different types of astrocytic glioma, such as glioblastoma and cerebellar astrocytoma, will both exhibit progressive gliofibrillogenesis (5). Indeed, one of the striking results that emanates from our studies, confirmatory of the classic tissue culture investigations by LUMSDEN (1), is the astrocytic nature of most of the cells that compose the glioblastoma. In this tumor, progressive differentiation to fibrillary astrocytes is quite apparent by both light and electron microscopic criteria. Neuroglial fibrils -- demonstrated in the light microscope with MALLORY's phosphotungstic acid/hematoxylin (PTAH) -- are usually conspicuous after 4 to 6 weeks in vitro (4), and an increase in cytoplasmic filaments (Fig. 5) is evident by electron microscopy after 3 to 4 weeks (5). In experimental ENU-induced rat cerebral and spinal astrocytomas, increased glial fiber formation was noted by light microscopy to occur after 35 to 40 days in culture, and a moderate increase in cytoplasmic filaments (Fig. 6) was seen by electron microscopy after 18 to 35 days (12). Both in the human and experimental gliomas, the filaments, which usually measure 80 to 90 Å in diameter, form broad bundles characteristic of glial filaments and are sometimes accompanied by an increase of closely associated granular endoplasmic reticulum and free ribosomes, a morphologic feature highly suggestive of increased protein synthesis. Protein synthesis is, in this case, presumably related to the increased production of glial fibrillary acidic (GFA) protein in the tumor cells. This protein has been shown to be the main constituent of astrocytic fibers in the central nervous system and can be used as a biochemical marker for fibrillary astrocytes (14-16). Dr. AMICO BIGNAMI, who has collaborated in some of our studies, has demonstrated by immunofluorescence the presence of GFA protein in the explants of several glioblastomas, two cerebral astrocytomas and two ENU-induced rat astrocytomas that had been maintained up to several weeks on sponge foam matrices.

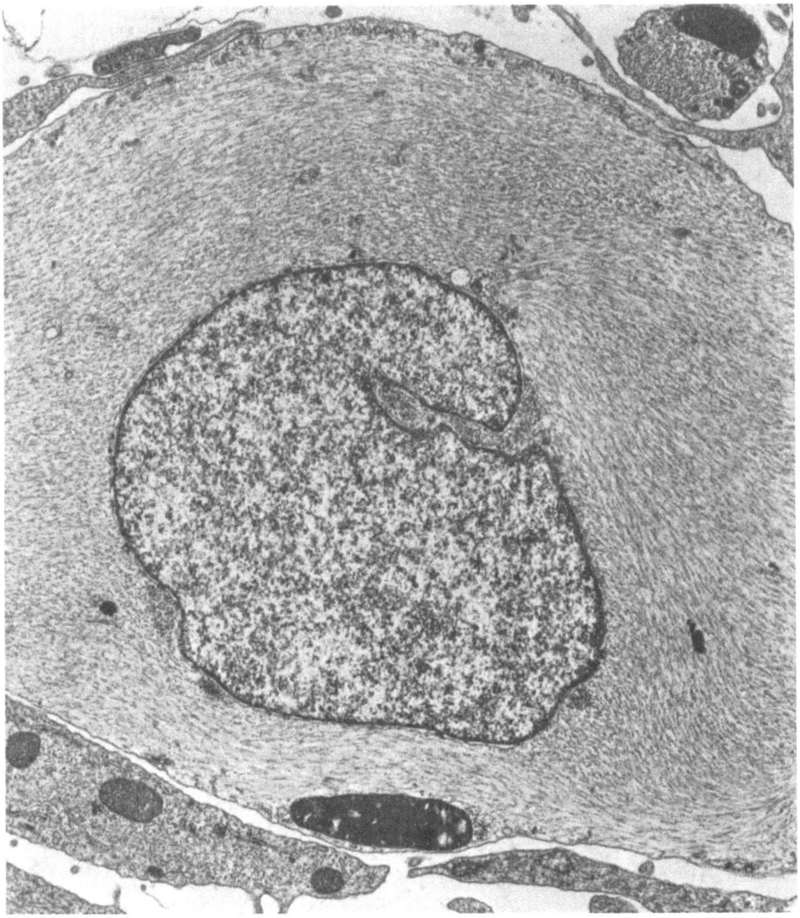

Fig. 5. Glioblastoma on sponge foam, 20 days in vitro. Tumor cell demonstrates extreme fibrillogenesis with displacement of the other organelles to the periphery. x 16,000

Increased gliofibrillogenesis in vitro as a morphologic expression of differentiation is not limited to gliomas of astrocytic origin. The same phenomenon was seen in two ependymomas. This finding is of course not unexpected in view of the well-established observation of cytoplasmic filaments in both normal and neoplastic ependymal cells (17-19). In both the spinal myxopapillary ependymoma (10) and the fourth ventricle cellular ependymoma (13) that we cultured, the glial nature of the filaments was confirmed by the presence of a strongly positive immunofluorescence reaction to GFA protein (Fig. 7). Another example of gliofibrillogenesis is provided by a remarkable case of malignant cerebellar and brain stem glioma in a 14-year-old girl that showed the concomitant picture of diffusely infiltrating astrocytoma and focal medulloblastoma. In this tumor, both tissue and organ culture demonstrated a progressive differentiation from primitive medulloblastoma cells into astrocytes rich in glial filaments (11).

Other in vitro systems used for the study of astrocytic gliomas have likewise been shown to favor gliofibrillogenesis (1), but this phenom-

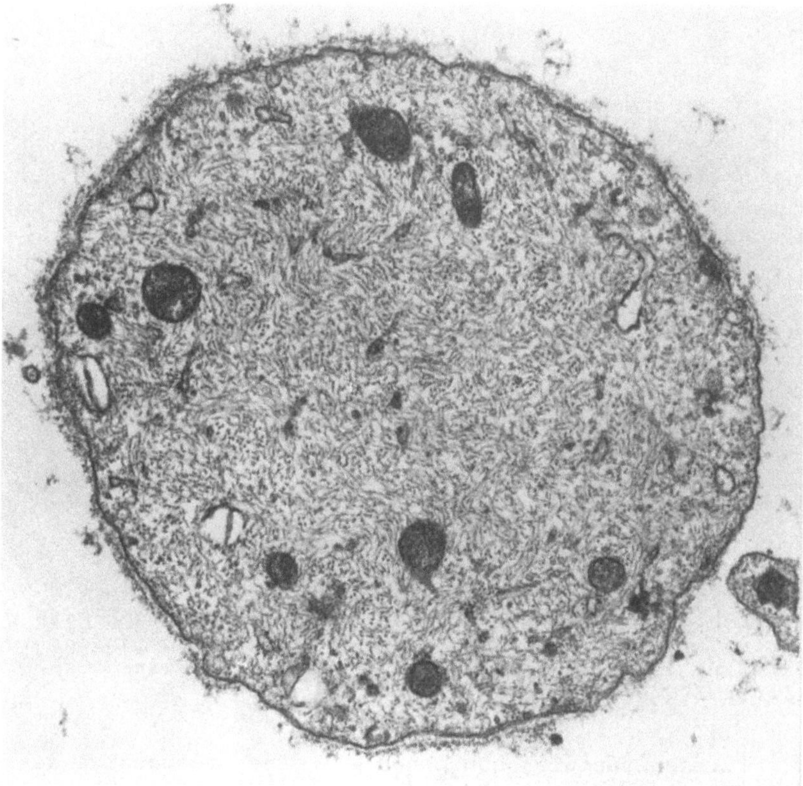

Fig. 6. Experimental ENU-induced rat spinal astrocytoma, 28 days in vitro. Note prominent filaments in tumor cell process, which is invested by a well-defined basement membrane. x 24,300

enon has not invariably been reported by others as a feature of gliomas maintained either as primary explants or in monolayer or three-dimensional matrix systems (see 5, for references). Moreover, many tissue and cell culture systems have long been known to exhibit an increased production of cytoplasmic fibrils ("stress fibers" or "tension striae") with increasing age of the cultures. These fibrils are usually regarded as a form of structural adaptation to tissue culture environment (20, 21). We do not know whether, or to what extent, the in vitro conditions of monolayer culture on glass surface are comparable to those of organ culture systems, but our belief that increased gliofibrillogenesis in our systems represents true differentiation and not a phenomenon primarily related to a cellular adaptation to tissue culture environment is based on the following data: 1) the strong affinity of the glial cell processes and intracellular fibrils for the PTAH stain; 2) the substructural configuration of the 80 to 90 Å cytoplasmic filaments, which, under high magnification, show in cross section four globular subunits measuring 21 Å in diameter connected by crossbars and thus forming tubular structures whose inner lumen measures 25 Å in diameter (5); 3) the invariably positive results obtained by immunofluorescence against the antigen of GFA protein; and 4) the occasional association of these filaments with the characteristic osmiophilic bodies known as Rosenthal fibers, which have been previously noted in astrocytic gliomas both in vivo (18) and in vitro (22).

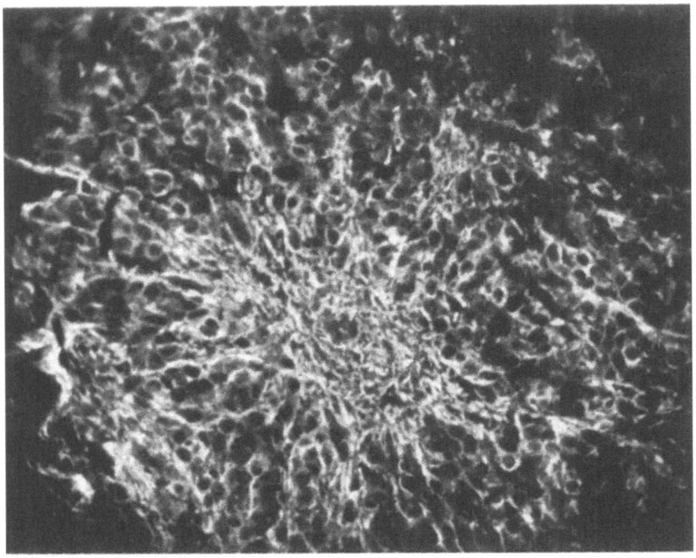

Fig. 7. Same tumor as Figs. 3 and 4, on sponge foam, 16 days in vitro. The cytoplasm and cell processes of this ependymoma give a strongly positive immunofluorescence reaction against the antigen of the glial fibrillary acidic protein. The nuclei are unstained. x 200. (Courtesy of Dr. AMICO BIGNAMI)

Three further features concerned with the progressive astrocytic differentiation of malignant gliomas in vitro should be commented on. The first is our observation that explants maintained in organ culture systems may demonstrate differentiating features that are in marked contrast with those of sister cultures grown on collagen-coated coverslips (4). We have noted that some of the latter may, after 4 to 6 weeks, show markedly anaplastic cytologic features while sister explants in organ culture demonstrate a much more differentiated pattern of highly fibrillated astrocytes. Such a difference emphasizes, as has long been known, the probable role played by the artificial conditions of culture in determining the morphologic character of human tumors in various in vitro systems.

Secondly, we draw attention to the contrast that exists between the general tendency of organ culture systems to favor differentiation, and the phenomenon of dedifferentiation (or anaplasia), which has been extensively described as occuring in astrocytic gliomas in vivo (18). Tissue culturists have inferred from their numerous observations on gliomas in traditional culture systems that progressive cellular dedifferentiation -- in the sense used by histopathologists -- does not play a role in primary explant cultures (1, 2). Our observations on human and experimental nervous system tumors maintained in organ culture systems so far support their conclusion. The paradoxical situation therefore exists in which a tumor presumed to have exhibited progressive anaplasia in vivo will reverse its cellular evolution along the pathway of differentiation when transferred to an in vitro system. This has long been known to be the case in glioblastoma. It was strikingly demonstrated in the case, already referred to, of cerebellar and brain stem glioma in which the concomitant picture of a diffusely infiltrating astrocytoma and focal medulloblastoma was present. In this tumor, the in vitro studies demonstrated progressive differentiation from medulloblastoma cells into astrocytoma cells, yet the clinical, gross and some of the microscopic features at autopsy dictated the alternative interpretation that the neoplasm was, in all probability, an exceptional example of diffuse cerebellar astrocytoma that had recently dedifferentiated into a medulloblastoma. The reasons for this interpretation and their implications in terms of the differentiating capabilities of the medulloblastoma have been detailed elsewhere (11).

Thirdly, recent observations in our laboratory suggest that astrocytic differentiation at the biochemical level may be demonstrable in organ culture systems in the absence of morphologically identifiable glial filaments. For the purpose of these experiments, we have used the C-6 line of rat glioma cells, which had been cloned from an N-nitrosomethylurea-induced experimental tumor (23, 24). The cells were obtained from the American Type Culture Collection and maintained as monolayer cultures. The line, which was originally selected on the basis of a high production of protein S-100, consists of relatively poorly differentiated glial cells that do not form bundles of glial filaments when grown in either monolayer culture, suspension culture, or organ culture on sponge foam. When grown in a three-dimensional matrix system, the histologic architecture of the C-6 glial cell line is similar to that seen in poorly differentiated human astrocytic gliomas and in some of the experimental ENU-induced rat astrocytomas we obtained in our laboratory and which were subsequently maintained in similar organ culture systems. By contrast, a marked increase of GFA protein is demonstrable by immunofluorescence after 12 to 14 days in organ culture, whereas only a very occasional positive cell is seen in cultures of the same cells maintained in monolayer or suspension cultures (25). Therefore, if synthesis of a protein specific for fibrillary astrocytes is to be regarded as a marker of differentiation, then it seems that, in these *in vitro* systems, the synthesis of GFA protein can proceed without the formation of intracytoplasmic glial filaments, which have until now been regarded as an important morphologic differentiating marker of astrocytes. As no gliofibrillogenesis could be demonstrated in the C-6 cells at any stage *in vitro*, the hypothesis that synthesis of GFA protein is an event that precedes the formation of glial filaments lacks a basis for support at this time. All that can be inferred so far is that, in the conditions studied, the C-6 rat glioma represents a still morphologically undifferentiated glial cell type with the potential to differentiate into fibrillary astrocytes, and that this potential is first demonstrated by the increased production of GFA protein. This could occur either non-specifically as part of a general increase in protein synthesis, or more specifically as the result of increased cell-to-cell contacts, as shown by the increase in S-100 protein in monolayer C-6 cultures. Further studies of this problem are now actively in progress, but preliminary experiments by Dr. LAWRENCE F. ENG on the increased production of GFA protein in rat C-6 glioma cells in organ culture have shown, by using a "two-site" immunoradiometric assay, a 10- to 20-fold increase of this protein compared to the levels found in monolayer or suspension cultures.

2. Ependymal Differentiation

A different aspect of differentiation was demonstrated in the sequential evolution of a human ependymoma in which the attention was specially focused on the *in vitro* development of ependymal rosettes (13). As shown in Fig. 3, the histologic organization of the tumor was very well maintained in organ culture. Well-defined ependymal rosettes were easily identified by light microscopy in virtually all the samples examined, up to the last date of observation, i.e., up to 86 days *in vitro*. Although it was impossible -- short of serially sectioning all the specimens -- to determine by light microscopy whether the number of ependymal rosettes had actually increased in culture, electron microscopy demonstrated, after 3 weeks, evidence of increasing differentiation, which consisted in an increasing prominence of long convoluted zonulae adherentes that alternated with focal elaborations of microvilli. Caution is of course needed in extrapolating from the static morphologic data provided by electron micrographs, but the impression was gained that a stage of "pre-rosette" formation developed during which compact surface infoldings resembling microvilli were formed discontinuously on the tumor cells. These complex infoldings often showed interconnections at sites where no lumens of rosettes could be discerned. As rosettes became apparent, some of the lumens were packed with very large numbers of closely apposed microvilli (Fig. 8). Thus rosettes seemed to develop as the result of the separation of groups of cells that were at first interconnected by unusually long and complex zonulae adherentes, while at the same time segments of the plasma membrane immediately adjacent to the zonulae developed complex infoldings with the morphologic features of microvilli. In newly formed rosettes, these microvilli filled the lumen almost completely. No definite temporal relationship could be inferred in relation to cilia-formation, as at least some cilia -- which contained variable numbers of paired or unpaired microtubules -- were invariably seen at all stages of rosette formation (Fig. 8).

Another feature that suggested increasing differentiation of the ependymoma *in vitro* was the frequent observation of a lining of columnar ependymal cells on the deep surface of the explants, facing the supporting foam matrix (Fig. 4). This phenomenon was almost invariably demonstrated in cultures more than 4 weeks old. By contrast, the exposed surface of the explants, i.e., at their gas-nutrient medium interface, was invariably lined by flattened undifferentiated cells (Fig. 4). Presumably, ependymal differentiation is dependent in part on the various nutritional conditions that are operative in different areas of the explants. Alternatively, the fluid medium adjacent to the solid explant may provide the lining tumor cells with a physicochemical environment that is more analogous to the usual *in vivo* condition, in which a single lining of ciliated ependymal cells is normally in immediate contact with the ventricular cerebrospinal fluid.

These observations, which suggest that increasing differentiation occurs in ependymoma maintained in organ culture, are essentially in agreement with those of KERSTING (2), according to which increasing numbers of rosettes can be demonstrated in the monolayer outgrowths of ependymomas cultured on glass. We have already mentioned above the other trend of differentiation that may be elicited in the ependymoma, although not along ependymal lines, namely progressive gliofibrillogenesis (see Fig. 7). Finally, we have noted an increased number of glycogen rosettes in this tumor, a feature that has long been known to be characteristic of astrocytic cells (13).

3. Vascular Stromatous Changes

Two features characteristic of glioblastomas *in vivo* have so far not been observed in our organ culture systems. The first is so-called pseudopalisading of the tumor cells. Its absence was surprising in view of its frequency *in vivo*. The second is vascular endothelial proliferation. On the contrary, the vascular stroma shows a marked tendency to undergo sclerosis, with thickening, hyalinization and gradual obliteration of the vascular lumens (4). By electron microscopy, the

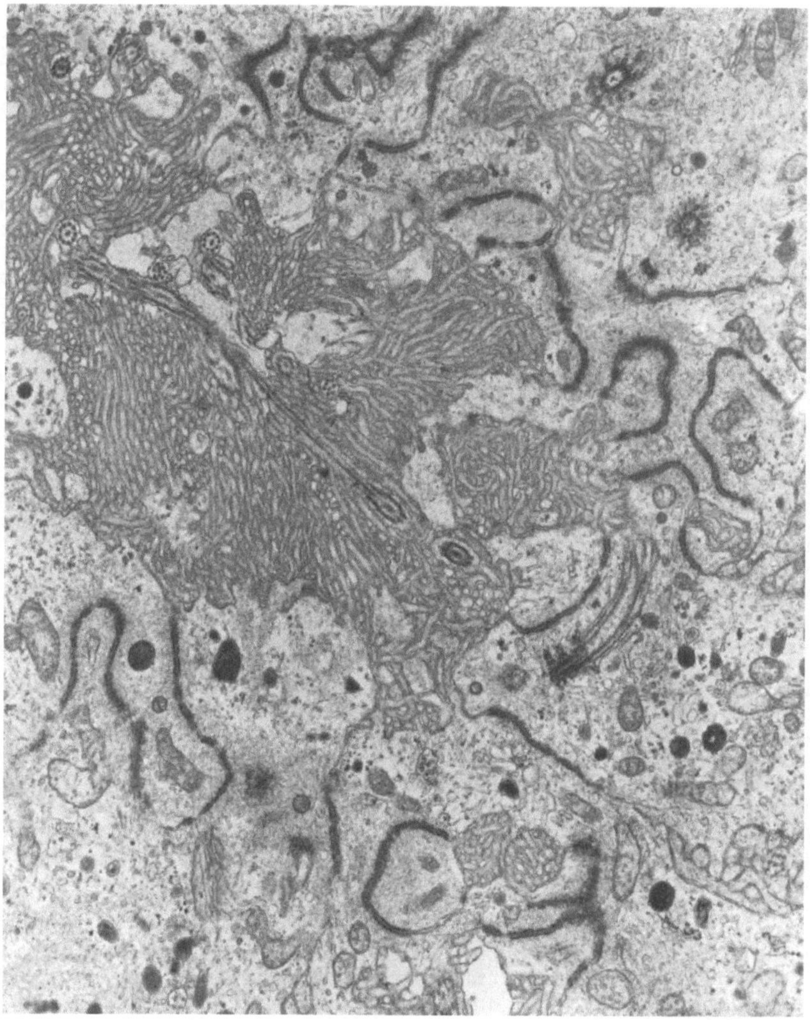

Fig. 8. Same tumor as Figs. 3, 4 and 7, cultured on sponge foam, 30 days in vitro. Note elaborate zonulae adherentes between the tumor cells, focal collections of microvilli and (left of center) one ependymal rosette packed with microvilli and containing several cilia. x 10,300

early stages of perivascular sclerosis are characterized by an increasing production of convoluted basement membrane material in the pericapillary and other perivascular spaces (5). This is often accompanied by increased collagen formation. In most cases, the increased connective tissue appears to be related to the endothelial cells or pericytes that have been explanted with the tumor. It is therefore not surprising that in the exceptional case of a glioma such as in the myxopapillary ependymoma of the filum terminale, in which there normally is an abundant supportive connective tissue stroma (19), the increased production of basement membrane material along the tumor cell surfaces was extremely conspicuous. As described and illustrated elsewhere (10), the basement membranes thus newly formed developed into elaborate convolutions that became separated from the tumor cell processes. This not only seemed to parallel an increase of collagen fibrils but also was apparently related to the formation of fine extracellular filaments adjacent to collagen fibrils and basement membrane, probably representing collagen fibril precursors.

The phenomenon of vascular sclerosis and hyalinization does not appear to be specific to gliomas, and we have seen it in other tumors, including capillary hemangioblastoma, acoustic Schwannoma, experimentally ENU-induced malignant neurilemomas, metastatic carcinoma, and pituitary adenoma. Possibly, the proliferation of vascular endothelium so characteristic of malignant gliomas may be contingent on factors found only in vivo, the most likely of which is the presence of a vascular circulation. It may be that the absence of a blood supply in organ culture may act as a stimulus to the proliferation of collagen fibrils and the increased production of basement membrane material in a manner analogous to the phenomenon of "involution" seen in blood vessels in vivo as the result of permanent arrest or closure of the local blood circulation. It is, however, clear that the proliferative capacity of the endothelial cells themselves remains potentially unimpaired in the artificial conditions of culture when they are neoplastic, as we have successfully maintained a capillary hemangioblastoma of the cerebellum up to 48 days in organ culture systems (7). It seems therefore that it is the neoplastic character of the cells that favors their maintenance and further growth in vitro. This character is presumably absent from the supporting vascular stromal cells found in malignant gliomas.

The increased production of basement membrane material appears, in all probability, to be mediated through endothelial cells or closely related cells of mesenchymal origin. These cells are always present in gliomas, irrespective of the histologic type of the tumor. As reported elsewhere, the increased basement membrane material thus formed is often found to be apposed to the plasma membranes of adjacent astrocytic (5) or ependymal (10) tumor cell processes, therefore accentuating a normal feature of gliovascular basement membrane relationship. It is indeed well known that in normal circumstances glial cells share a basement membrane which is related to the adjacent mesenchymal tissue such as vascular endothelium (26, 27). This feature is especially evident in the myxopapillary ependymoma of the lower end of the spinal cord (10, 19), in which the presence of large amounts of basement membrane material in vivo and its excessive proliferation in vitro are presumably attributable to the normal anatomic relationship of the ependymal cells in the conus medullaris and filum terminale, where these cells are directly apposed to connective tissue derived from the leptomeninges instead of resting upon the neuropil. A similar increase of basement membrane material surrounding the processes of astrocytic tumor cells has been noted by us in the later stages of culture of ENU-induced experimental rat astrocytomas (Fig. 6) (12).

On the other hand, we cannot exclude the possible role of cells other than fibroblastic (or other mesenchymal) cells through which the production of increased connective tissue material might be mediated. It is known that normal and neoplastic epithelial cells are capable of forming basement membranes and that embryonal epithelium and neuroepithelium may, in certain circumstances, be responsible for the synthesis of both basement membrane material and collagen fibrils (see 10, for references). It has also been shown that in cystic degeneration of senile human eyes the Müller cells from the retina, which are neuroglial, may be the source of collagen or collagen-like fibrils and of increased basement membrane material (28), and it is of course well known that neuroglial elements of the normal mammalian brain exhibit basement membrane material as a form of surface specialization when they are in contact with collagen (see 19, for references). This presumably is a specialized form of cellular differentiation. The possibility that in organ culture systems there may be physicochemical mechanisms that favor the increased production of both collagen and basement membrane material by cells of nonmesenchymal origin should therefore be raised. Since some epithelial and mesenchymal basement membranes are known to share the same antigenic components, could the same factors that mediate the increase of collagen fibrils in vitro also be responsible for the increased production of basement membrane material by neuroepithelial cells? Studies using amino-acid precursors for the investigation of collagen protein synthesis are obviously needed to help resolve this interesting problem.

SUMMARY AND CONCLUSIONS

The feasibility of organotypic culture methods for the growth and maintenance of human and experimental gliomas has been demonstrated, using both a three-dimensional sponge foam matrix technique and a Millipore filter platform technique. Cultures can generally be maintained from one to two months, and up to four months in the case of a glioblastoma. The patterns of histologic organization and differentiation seen in vivo are faithfully maintained. Increasing cellular differentiation occurs irrespective of the anaplastic character of the original tumors and is a general feature of this system. The essential astrocytic nature of most human glioblastomas is confirmed. The technique is suitable for the study of specific problems of histogenesis and differentiation, and has given so far interesting data on gliofibrillogenesis as an in-vitro phenomenon, using both biochemical and morphologic markers. We have applied it to study the genesis of rosette formation in an ependymoma and to elucidate the differentiating capabilities of poorly differentiated experimentally induced gliomas, e.g., the C-6 rat glioma. Increased glial fibrillogenesis is accompanied by the production of antigenically demonstrable glial fibrillary acidic (GFA) protein, a specific marker for fibrillary astrocytes, and by morphologic evidence suggesting increased protein synthesis. However, increased production of GFA protein may also occur in the absence of gliofibrillogenesis. The phenomenon of differentiation in vitro stands in contrast to that of dedifferentiation, which is characteristic of many human gliomas in vivo.

In organotypic systems, successful maintenance in vitro preferentially selects neoplastic cells, as opposed to the supporting vascular elements, which undergo degeneration. A typical feature is the development of perivascular sclerosis, leading to the progressive obliteration of vascular lumens and corresponding, by electron microscopy, to a conspicuous increase of basement membrane material and collagen fibrils. The increased basement membrane material often preserves a close relationship to the plasma membranes of the adjacent glial tumor cells and, although the production of fibrous connective tissue is in all likelihood mediated by explanted mesenchymal cells, the possibility that in these systems it might also be mediated by non-mesenchymal cells should be investigated.

Acknowledgements

We acknowledge the participation in this work of the following Fellows in our neuropathology training program: Drs. F. K. CONLEY, D. G. RAWLINSON, J. C. SIPE, A. M. SPENCE, and J. VRAA-JENSEN. We thank Dr. AMICO BIGNAMI for the immunofluorescence studies, and Miss VIRGINIA L. FOLEY, Miss CAROL INESON, Mr. ROBERT KITCHEN, Mrs. MARY ANN LAWRENCE and Mrs. ELIZABETH VANEK for technical assistance. We also express our gratitude to our neurosurgical colleagues, especially Drs. J. W. HANBERY and J. GOLDEN, for their unfailing cooperation in providing the human material for these studies.

REFERENCES

1. LUMSDEN, C. E.: The study by tissue culture of tumours of the nervous system. In: Pathology of Tumours of the Nervous System. 3rd ed., p. 334. D. S. RUSSELL, and L. J. RUBINSTEIN (Eds.). London: Edward Arnold Ltd. 1971.
2. KERSTING, G.: Tissue culture of human gliomas. In: Progress in Neurological Surgery. H. KRAYENBÜHL, P. E. MASPES, and W. H. SWEET (Eds.) vol. 2, p. 165. Chicago: Yearbook Medical Publishers 1968.
3. SCHARENBERG, K., LISS, L.: Neuroectodermal Tumors of the Central and Peripheral Nervous System, p. 183. Baltimore: Williams and Wilkins 1969.
4. RUBINSTEIN, L. J., HERMAN, M. M., FOLEY, V. L.: In vitro characteristics of human glioblastomas maintained in organ culture systems. Light microscopy observations. Am. J. Path. 71, 61 (1973).
5. SIPE, J. C., HERMAN, M. M., RUBINSTEIN, L. J.: Electron microscopic observations on human glioblastomas and astrocytomas maintained in organ culture systems. Am. J. Path. 73, 589 (1973).
6. RUBINSTEIN, L. J., CONLEY, F. K., HERMAN, M. M.: Studies on experimental malignant nerve sheath tumors maintained in organ culture systems. I. Light microscopy observations. In preparation.
7. SPENCE, A. M., RUBINSTEIN, L. J.: Cerebellar capillary hemangioblastoma. Its histogenesis studied by organ culture and electron microscopy. Cancer, in press (1975).
8. CONLEY, F. K., RUBINSTEIN, L. J., SPENCE, A. M.: Studied on experimental malignant nerve sheath tumors maintained in organ culture systems. II. Electron microscopy observations. In preparation.
9. SPENCE, A. M., RUBINSTEIN, L. J., CONLEY, F. K., HERMAN, M. M.: Studies on experimental malignant nerve sheath tumors maintained in organ culture systems. III. Melanin pigment and melanogenesis in experimental neurogenic tumors: a reappraisal of the histogenesis of pigmented nerve sheath tumors. In preparation.
10. RAWLINSON, D. G., RUBINSTEIN, L. J., HERMAN, M. M.: In vitro characteristics of a myxopapillary ependymoma of the filum terminale maintained in tissue and organ culture systems. Light and electron microscopic observations. Acta neuropath. (Berl.) 27, 185 (1974).
11. RUBINSTEIN, L. J. HERMAN, M. M., HANBERY, J. W.: The relationship between differentiating medulloblastoma and dedifferentiating diffuse cerebellar astrocytoma. Light, electron microscopic, tissue, and organ culture observations. Cancer 33, 675 (1974).
12. SIPE, J. C., RUBINSTEIN, L. J., HERMAN, M. M., BIGNAMI, A.: Ethylnitrosourea-induced astrocytomas: morphologic observations on rat tumors maintained in tissue and organ culture systems. Lab. Invest. 31, 571 (1974).
13. VRAA-JENSEN, J., HERMAN, M. M., RUBINSTEIN, L. J., BIGNAMI, A.: In vitro characteristics of a fourth ventricle ependymoma maintained

in organ culture systems. Light and electron microscopic observations. To be submitted.
14. ENG, L. F., VANDERHAEGEN, J. J., BIGNAMI, A., GERSTL, B.: An acidic protein isolated from fibrous astrocytes. Brain Res. 28, 351 (1971).
15. BIGNAMI, A., ENG, L. F., DAHL, D., UYEDA, C. T.: Localization of the glial fibrillary acidic protein in astrocytes by immunofluorescence. Brain Res. 43, 429 (1972).
16. BIGNAMI, A., DAHL, D.: Astrocyte-specific protein and neuroglial differentiation. An immunofluorescence study with antibodies to the glial fibrillary acidic protein. J. Comp. Neurol. 153, 27 (1974).
17. BRIGHTMAN, M. W., PALAY, S. L.: The fine structure of ependyma in the brain of the rat. J. Cell Biol. 19, 415 (1963).
18. RUSSELL, D. S., RUBINSTEIN, L. J.: Pathology of Tumours of the Nervous System, 3rd ed. 122, 126, 156. London: Edward Arnold (1971).
19. RAWLINSON, D. G., HERMAN, M. M., RUBINSTEIN, L. J.: The fine structure of a myxopapillary ependymoma of the filum terminale. Acta neuropath. (Berl.) 25, 1 (1973).
20. LEWIS, W. H., LEWIS, M. R.: Behavior of cells in tissue culture. In: General Cytology, p. 385-447, edited by E. V. COWDRY. Chicago: University of Chicago Press 1924.
21. BUCKLEY, I. K., PORTER, K. R.: Cytoplasmic fibrils in living cultured cells. Protoplasma 64, 349 (1967).
22. GULLOTTA, F., FLIEDNER, E.: Spongioblastomas, astrocytomas and Rosenthal fibers: ultrastructural, tissue culture and enzyme histochemical investigations. Acta neuropath. (Berl.) 22, 68 (1972).
23. BENDA, P., LIGHTBODY, J., SATO, G., LEVINE, L., SWEET, W.: Differentiated rat glial cell strain in tissue culture. Science 161, 370 (1968).
24. BENDA, P., SOMEDA, K., MESSER, J., SWEET, W. H.: Morphological and immunochemical studies of rat glial tumors and clonal strains propagated in culture. J. Neurosurg. 34, 310 (1971).
25. BISSELL, M. G., RUBINSTEIN, L. J., BIGNAMI, A., HERMAN, M. M.: Characteristics of the rat C-6 glioma maintained in organ culture systems. Production of glial fibrillary acidic protein in the absence of gliofibrillogenesis. Brain Res. 82, 77 (1974).
26. PETERS, A., PALAY, S. L., WEBSTER, H. deF.: The Fine Structure of the Nervous System. The Cells and their Processes, p. 110. New York: Hoeber, Harper and Row 1970.
27. CALEY, D. W., MAXWELL, D. S.: Development of the blood vessels and extracellular spaces during postnatal maturation of rat cerebral cortex. J. Comp. Neurol. 138, 31 (1970).
28. GÄRTNER, J.: Extracellular fibril formation by neuroglial cells in cystoid degenerations of the peripheral retina of the human eye. Acta neuropath. (Berl.) 26, 45 (1973).

Scanning Electron Microscopy of Cerebral Tumors and Glial Cells

A. Arnold and D. Burrows

SUMMARY

The scanning electron microscope was used to study brain tumors and glial cells. The ependymal glial cells and astroglial cells were readily detected and compared with their morphology in light microscopy and phase microscopy. Similar studies were made with meningiomas. The beautiful and remarkable three dimensional views of these tumors obtainable with scanning electron microscopy was demonstrated. The techniques which we developed and utilized for the study of brain tumors were presented.

INTRODUCTION

The study of cerebral tumors and glial cells by scanning electron microscopy offers the researcher an opportunity to observe tumor morphology and tumor cell relationships in a three dimensional manner at both low and high magnifications. The conventional electron microscope has given us considerable information concerning the internal structure of cells as well as some knowledge of cellular membranes, but it is limited in its ability to give rise to any three dimensional images. The everyday study and visualization of cerebral tumors by routine light microscopy is fundamental to our clinical detection of these lesions, but a detailed study of these tumors is greatly restricted by the low limits of magnification and resolution of the optics in the light microscope. In our program cerebral tumors were first prepared for study in their natural relationships in the patient and then the tumor tissues were dissected out and dissociated for a study of their cellular and vascular constituents. Portions of the tumors were also taken for tissue culture and light microscopy. A comparison was then made of the morphology as observed by light microscopy, phase microscopy, and scanning microscopy.

TECHNIQUES

Patients with brain tumors were operated upon through the operating microscope. By use of the operating microscope a more careful selection of areas of the tumor could be secured for study. Portions of the tumor were taken for tissue culture, light microscopy, and scanning microscopy. The tissues selected for scanning microscopy were fixed in buffered glutaraldehyde and further examined under a stereomicroscope with magnifications from 50x to 200x. The tissue fragments could be teased apart under the stereomicroscope to allow a better exposure of cell surface and cell relationships to the scanning electron beam. Following this preliminary preparation the fragments of tumor were dehydrated through a graded series of alcohol to absolute alcohol. From absolute alcohol the tumor was critical point dried with carbon dioxide. The critical point dried specimens were carefully arranged on aluminum stubs and coated with gold or gold palladium in a vacuum coater. These prepared specimens were then viewed in a scanning electron microscope, with a field emission source of electrons at magnifications of 10x to 140,000x. Some portions of tumors were not coated with gold or gold-palladium, but simply were fixed in buffered glutaraldehyde, dehydrated through absolute alcohol, critical point dried, and then examined in the field emission type of scanning electron microscope. It is possible to examine uncoated cells and tissues in the field emission type of scanning electron microscope, since the beam of electrons in a field emission scanning electron microscope is many hundred times brighter than the beam of electrons produced by the thermionic tip of the more customary scanning electron microscope. This increase in beam brightness, in the field emission scanning unit, allows us to study the tissue or cells at very low accelerating voltages, below 1000 electron volts, and thereby avoid damage to the specimen from charging of the surfaces, which will occur when uncoated biological specimens are examined in the standard scanning electron microscope. Comparative studies were made of cerebral tumors, coated and uncoated, to detect any morphological changes produced by metallic coating.

The portions of tumor tissue removed for light microscopy were prepared by formaldehyde fixation, dehydration through a graded alcohol series to xylene, then imbedded in paraffin, and sectioned at 10u. Routine hematoxylin and eosin and various special staining techniques were utilized. These routine sections served as a basis for comparison for the cell and tumor details, as observed by the scanning microscope and in tissue culture.

The portions of tumor taken at surgery for tissue culture were carefully washed to remove excess blood and then fragmented. Falcon flasks containing Eagle's basal medium with Earle's salts and glutamine were seeded with the tumor. An antibiotic containing penicillin 10,000 units per ml. and streptomycin 10,000 micrograms per ml. was added to the medium. The flasks were incubated and observed with phase optics. The tumor cells were grown in monolayers. Some of the cell layers grown in the flasks were fixed <u>in situ</u>, dehydrated and stained by H & E, PTAH, and Cajal methods and examined. Other flasks with cultured cells were prepared for scanning microscopy by cutting out segments of the flasks on which the cells were growing. These were processed by glutaraldehyde fixation, alcohol dehydration, critical point drying with carbon dioxide, coating with gold or gold-palladium and then examined in the scanning electron microscope.

Fresh fragments of tumor were also dissociated in trypsin 0.25 070. The dissociated cells were fixed and collected on millipore filters or plastic surfaces. These preparations were dehydrated through graded alcohol, critical point dried with carbon dioxide, coated with gold or gold-palladium, and then examined in the scanning electron microscope.

In addition to the scanning electron mode, the scanning electron microscope allows for a scanning transmission mode. Thus it becomes possible to scan a tumor or its cells with the scanning beam, and also to view concurrently the internal detail of the cells with the transmission mode.

OBSERVATIONS

In Figures 1, 2, and 3 the three dimensional appearance of an ependymoma can be seen at magnifications of 1200x to 5000x. In this tumor the matrix is somewhat loose and therefore scanning images of cell detail are readily obtained. Small cavities can be seen between groups of tumor cells and blood cells. These small cavities are lined with the tumor ependymal cells and represent the rosettes seen in routine sections by light microscopy. The individual tumor cells appear spherical or polyhedral in shape. By contrast these cells when observed in a paraffin section by light microscopy are polygonal in shape since they are being viewed in single plane. The individual ependymal cells, as seen in light microscopy, suggest the presence of a long cytoplasmic process attached to blood vessels, which is difficult to see in any adequate detail. With the scanning microscope the cellular morphology is beautifully demonstrated. Under the scanning microscope the surfaces of the ependymal cells have numerous microvilli and infoldings. Some of the cells seen in the tumor are more elongated and cone like, with a blunted tip at the vertex of the cone. These cells and the more spherical cells possess elongated cytoplasmic processes, which may end in a fine tip or a foot like process (Fig. 4). One or two thick processes have been observed in some cells. With light microscopy the foot of the glial processes is not readily visible. The more slender ependymal cells make a gentle contact with external surfaces of small blood vessels. This type of arrangement of the ependymal cell to the vessel wall forms the perivascular rosette that we see in light microscopy and now in three dimension with the scanning electron microscope. In the rosette pattern, the nucleus lies a distance away from the vessel wall, and thin cytoplasmic processes extend from the cell and make very fine contacts with the vessel wall.

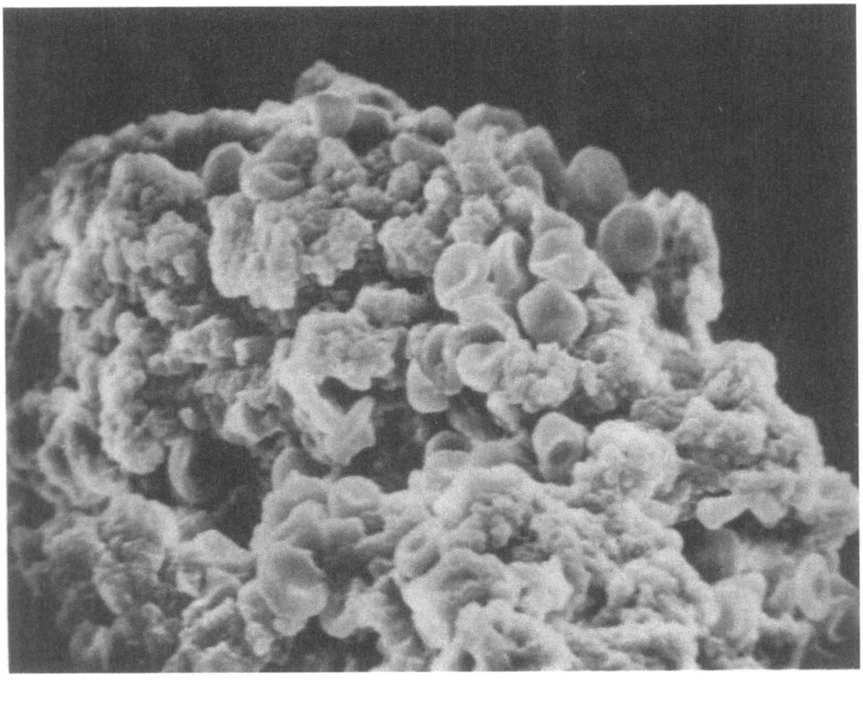

Fig. 1

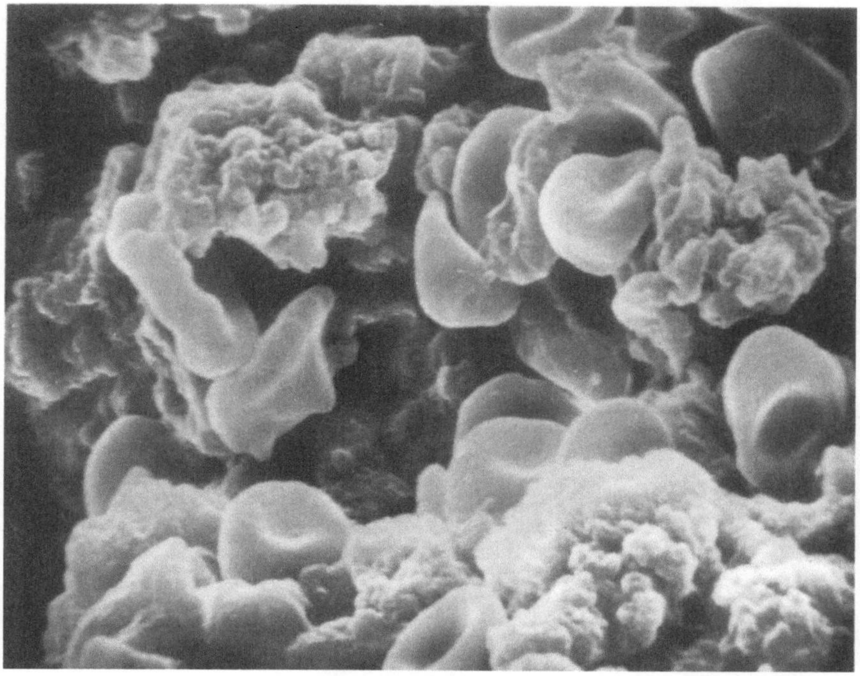
Fig. 2

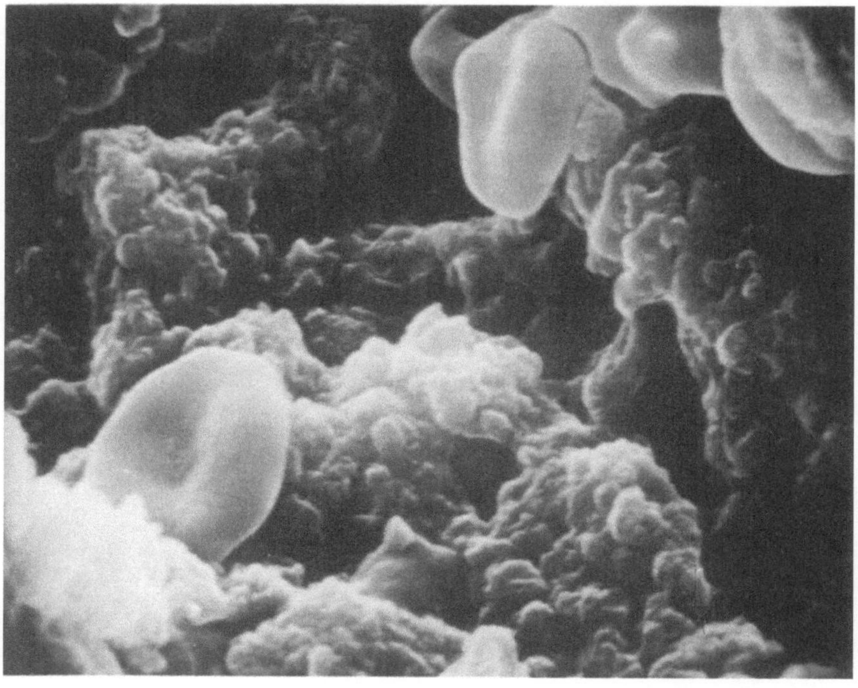
Fig. 3

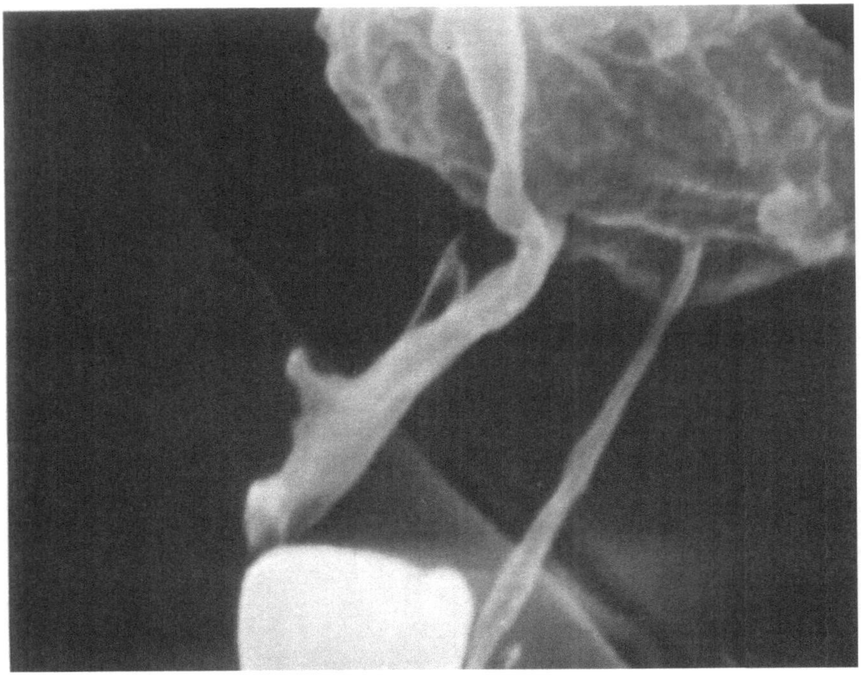

Fig. 4

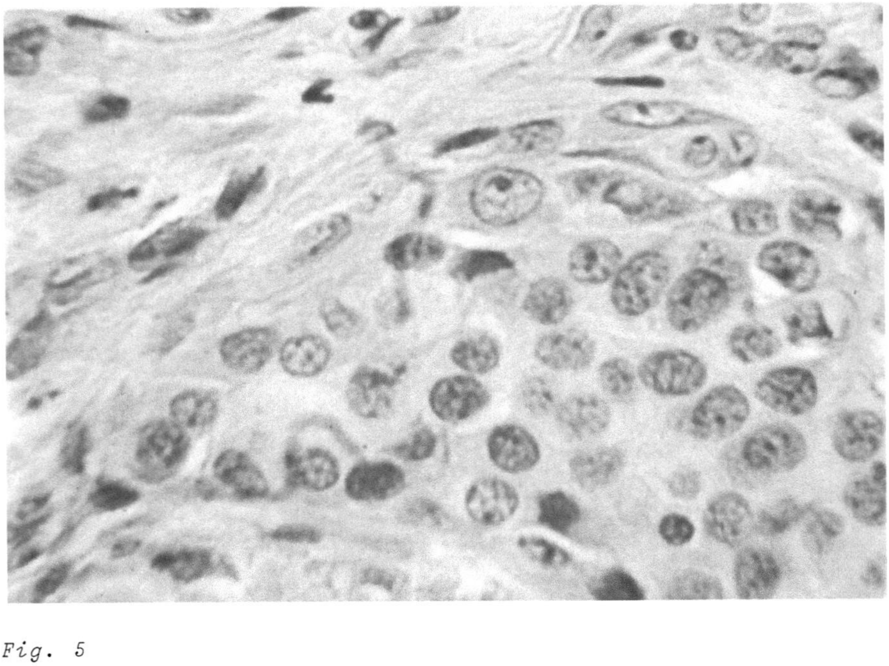

Fig. 5

The blood vessel components of cerebral tumors are readily studied by scanning methods. The surface detail of the vessels of all sizes, normal as well as abnormal, can be studied. In the ependymoma the walls of the micro vessels are usually quite regular and normal. The endothelial cells in these fine vessels can be seen bulging through the wall. In contrast the microvessels of the malignant astrocytoma and glioblastoma are irregular in contour, and the endothelial cells are irregular in size and shape. This abnormality of the endothelial cells of the malignant astrocytoma can be readily seen in light microscopy. Mitotic activity of the endothelial cell appears occasionally.

Fig. 5 illustrates the appearance of the meningotheliomatous type of meningioma in a paraffin section stained with hematoxylin and eosin, and viewed with the light microscope. In this type of meningioma, the cells are arranged in whorls or portions of whorls, depending on the plane of section. The cells are quite large, and possess one or more distinct and separate nuclei arranged in a syncytium. The cell outlines are indistinct in light microscopy. In the scanning image of the tumor the arrangement of cells is quite distinct (Fig. 6). The cells are quite large and appear to fuse with one another confirming the syncytial structure of the cell cytoplasm and multiple nuclei. The cell surfaces of the meningotheliomatous meningioma possess finer microvilli (Fig. 7) than the microvilli of the ependymal cells and astrocytes.

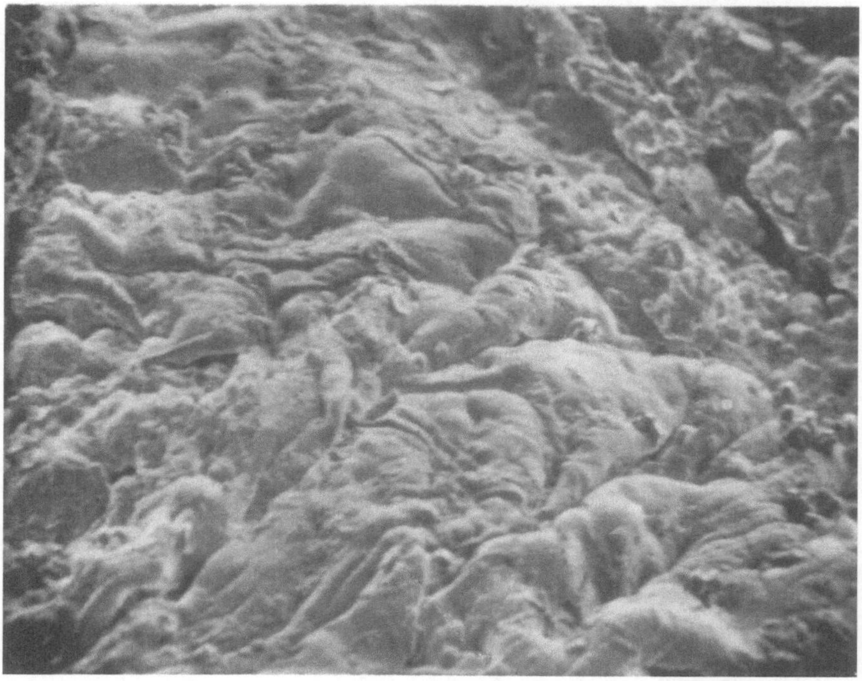

Fig. 6

Scanning microscopic views of a malignant astrocytoma are shown in Figs. 8, 9, and 10. The astrocytic cells, with their numerous processes are readily seen in a three dimensional view. The cell surfaces are covered with numerous microvilli. These microvilli vary in length and

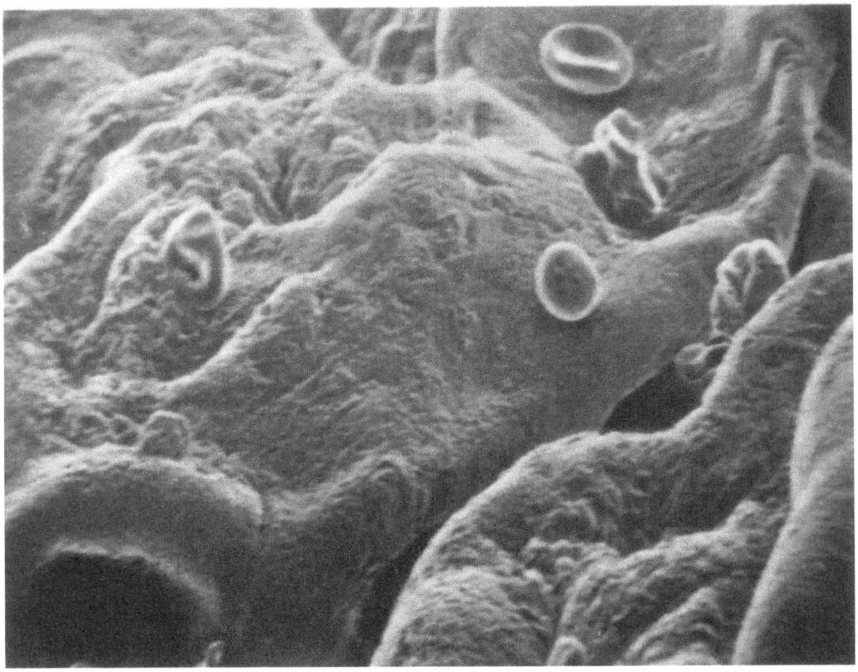

Fig. 7

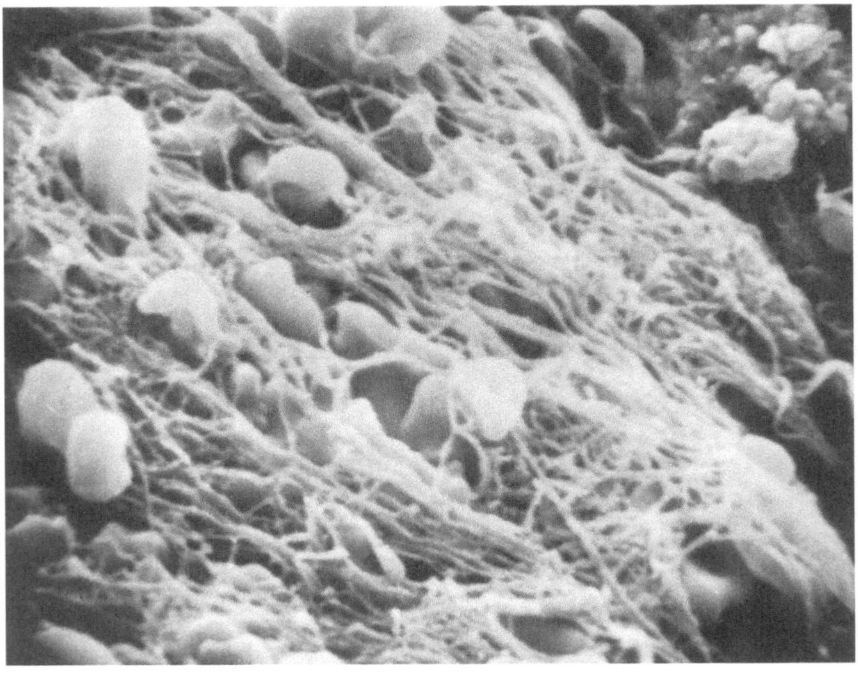

Fig. 8

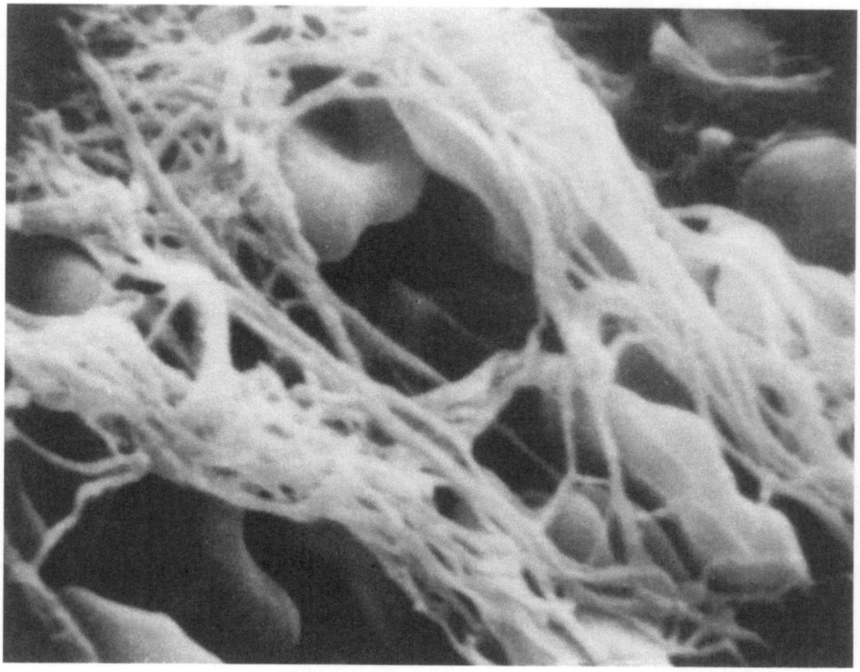

Fig. 9

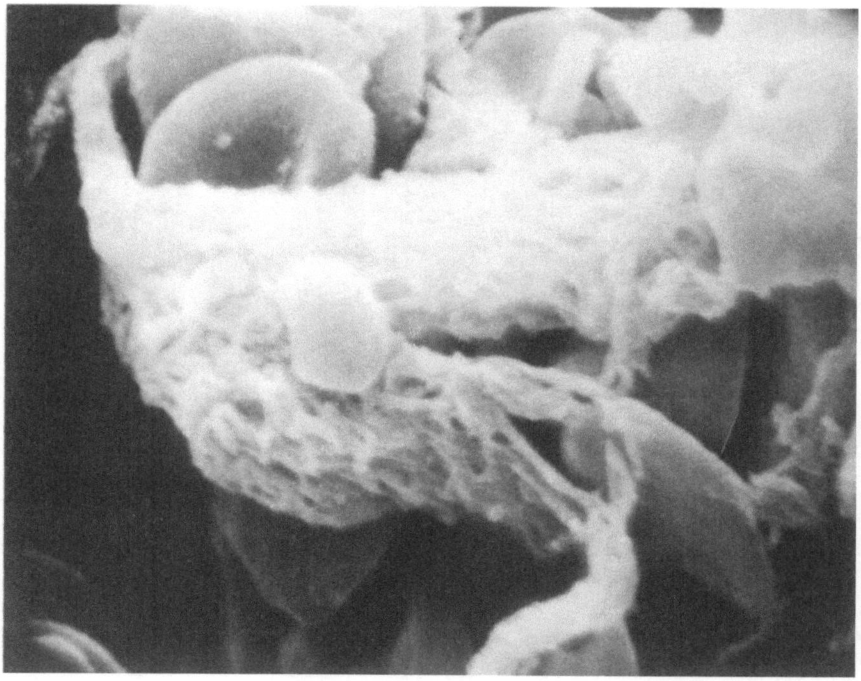

Fig. 10

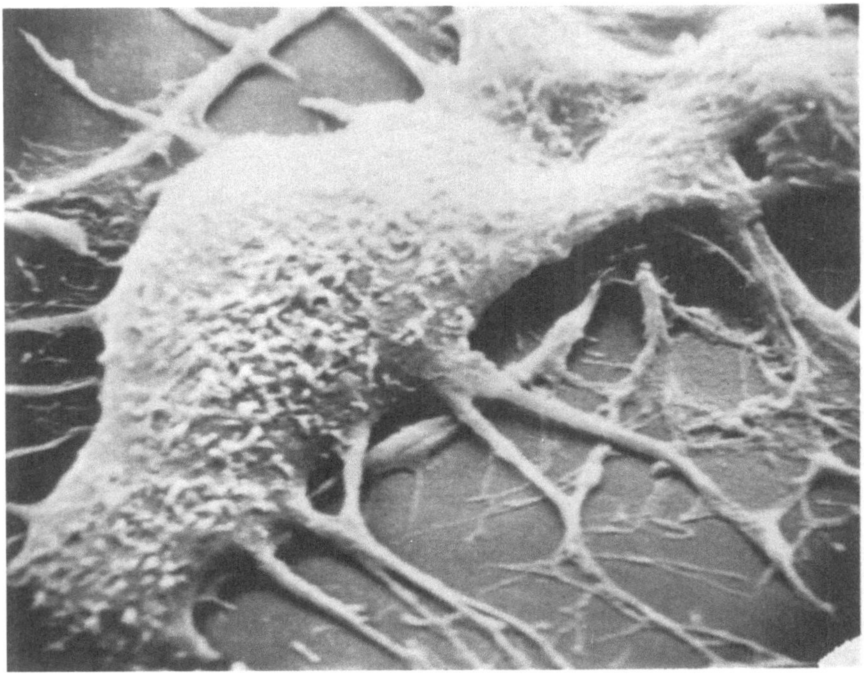

Fig. 11

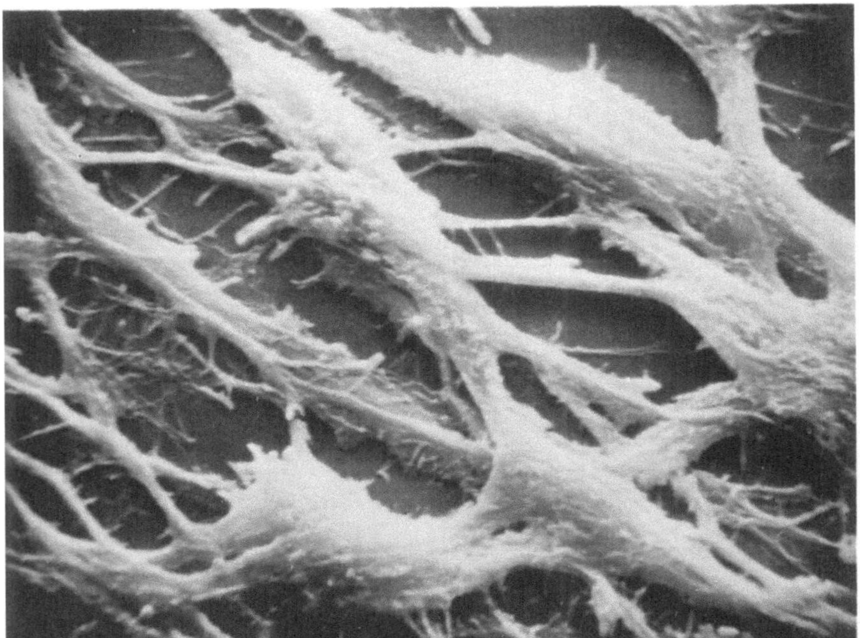

Fig. 12

thickness. The microvilli of the astrocyte are far more numerous and heavier than the microvilli of the ependymal cells. Astrocytic glial cells in tissue culture are quite comparable in appearance to the astrocytes seen in the tumor prior to culture (Fig. 11). The density of glial processes is quite striking when observed in culture (Fig. 12, 13). Obviously the vascular stroma, present in the primary tumor from the patient, occupies a considerable portion of the parenchyma of the malignant astrocytoma and the glioblastoma multiforme.

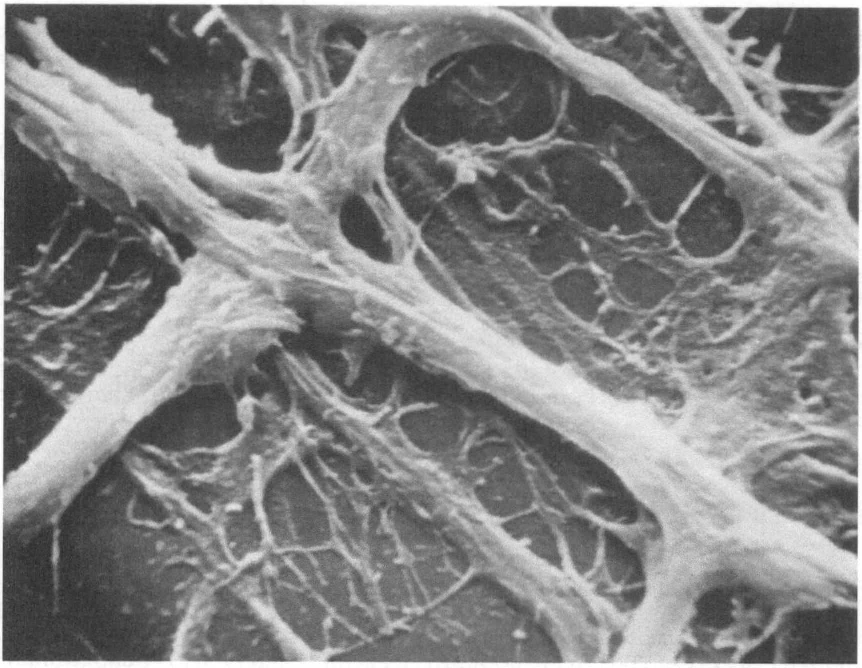

Fig. 13

DISCUSSION

Scanning microscopy was first used extensively in the study of flat surfaces of metals and minerals. Only in the last three or four years has the instrument been turned to biological materials (1, 2, 3). The early biological studies dealt with plant pollens, blood elements, sperm cells, and other dissociated biological materials or single cell elements. Our interest has been to literally scan the intact brain tumor fragment for its structural arrangement and cellular detail and then to compare our observations with our knowledge of brain tumors as seen by light microscopy. The dissociation of the tumor with trypsin is helpful to gain access to individual cells, especially when combined with collecting of the cells on millipore filters.

With a standard stereomicroscope the depth of field rapidly drops off as we approach a magnification of 200x. This is much too low for a study of cell detail. The light microscope with the finest optics offers very little depth of field and the resolution is limited to one-half micron. The scanning microscope offers a considerable depth of field, a very wide range of magnification from 10x to 100,000x or more, and a resolution approaching 50 angstroms. What is most important is the ability of the scanning beam to see the biological structure in its true three dimensional form. The surface detail of the cell or tumor can be characterized and identified. Alterations of the normal or abnormal cell, produced by chemical means or radiation, are readily detected by scanning images. When scanning electron microscopy is combined with a scanning transmission electron mode a complete characterization of the cell, externally and internally, is possible and will enhance our knowledge of the cellular components and composition of tumors of the central nervous system. In the transmission scanning electron mode the beam of electrons, under proper accelerating voltage, penetrates the specimen and is collected by a secondary detector placed beneath the sample. In tissue work thicker and larger pieces of tissue can be examined with greater ease with the transmission scanning electron microscope than with the standard transmission electron microscope. Examination of the specimen at very low magnifications is possible with the scanning electron and with the transmission scanning mode. This is very useful for orientation. The magnification can be readily raised without loss of orientation. With the use of visual storage monitors observations, made at various magnification levels or in different areas of interest, can be stored on the video screen and comparisons made as needed. Techniques now available can be used to process the signals and video images from the scanning electron microscope. These images can be readily stored on tape and analyzed or displayed. Thus it now appears possible to study and to completely characterize a tumor and its cellular constituents through use of the scanning electron microscope.

REFERENCES

1. LEWIS, E. R.: Studying neuronal architecture and organization with the S. E. M. p. 281-288. Proceedings 4th. Annual Scanning Electron Microscope Symposium 1971.
2. LODIN, Z., FALTEN, J., BOOKER, J., HARTMANN, J., SENSENBRENNER, M.: Fiber formation and myelinization of cultivated dissociated neurons from the chicken dorsal root ganglia - An E. M. and S. E. M. study. Neurobiology 3, 166 (1973).
3. VESELY, P., BOYDE, A.: The significance of S. E. M. evaluation of the cell surface of tumor cell biology, p. 689-696. Proceedings 6th. Annual Scanning Electron Microscope Symposium 1973.

Discussion

N. A. Vick

Drs. ZIMMERMAN, RUBINSTEIN and BIGNER have dealt explicitly and implicitly with almost all the important issues and problems in the biology of gliomas. At the same time, an underlying theme is apparent. This is their well-developed appreciation of the magnitude of the clinical problem, obvious despite their concern with issues requiring sophisticated laboratory methods.

At present, with experimental work progressing in many directions we still lack proof that human brain tumors are caused by any of the carcinogenic agents in use. Most would agree that the overriding issue of importance in brain tumor research is the quest for the discovery of the cause, or the causes, of the neoplastic transformation. Lacking this, it seems that we can nevertheless identify at least four issues of importance as suggested by the papers given by these experienced workers: 1, The concern in the past with classification of the gliomas on microscopic grounds may not be as relevant as previously considered. The various carcinogens, ranging from the aromatic hydrocarbons to the nitrosoureas to viruses may induce gliomas of quite similar structure. Taken as a group, they closely mimic the entire spectrum of neoplasia of which neuroectodermal cells are capable in man. 2, It can be assumed, from the wide range of species which can be utilized for the experimental production of brain tumors, that any or all of these carcinogens could or even may cause brain tumors in man. 3, The question of "dedifferentiation" and "differentiation", which has been of historical importance, may not be as central to our understanding of the etiology of gliomas as has been thought. It seems, as Dr. Rubinstein's organ culture work indicates, that a highly anaplastic glial tumor has the capacity to express at the subcellular level many of the characteristics of normal glial development. This serves to emphasize again that what we call a neoplastically transformed cell has lost only part of its destiny and suggests that these cells are capable of resumption of a more orderly state of being if only we knew how the control mechanisms had been altered. It is hard to know, of course, just how closely parallels can be drawn between that which occurs in culture, even organ culture and that which occurs in vivo. Organ culture allows observation of some basic phenomena characteristic of neuroectodermal cells. But it would seem that these are epiphenomena and that they are not necessary, in the philosophical sense, to the neoplastic transformation. All that can be said is that neoplastic neuroectodermal cells do seem to have the capacity to proceed to a state of development which mimics the normal cellular state of affairs. 4, Autochthonous experimental tumors, with a blood supply appropriated from the pre-existing normal brain parenchymal vessels, are probably the optimal experimental model for the evaluation of drugs potentially useful in the clinical setting. Dr. BIGNER's systematic approach with the avian sarcoma virus system, based on production of uniform cell-cultured virus of high titer, is confirming its obvious early promise. There would seem to be no doubt that this system is the best yet available for the evaluation of chemotherapeutic drugs of possible use in man despite the fact that we do not yet fully understand the mechanisms by which viruses transform cells.

New techniques will be applied with increasing accuracy to problems in brain tumor biology. The range of use of scanning electron microscopy, and its ultimate value will depend not only on further experience, but on close correlation of its images with those obtained by older and established methods. Dr. ARNOLD has provided us with some insight into the possible use of scanning electron microscopy; it remains to be seen whether the concept of surface and membrane changes in neoplasia (as this technique holds promise of revealing) will be as important as they seem now. Purely structural techniques, however, will always remain adjuncts to advances in tumor biology despite their obvious power and elegance. Structural investigations, by themselves, have never solved any important problems in the biology of neoplasia as is evidenced by the last decade or so of work with the transmission electron microscope.

Session II

Neuroradiologic Aspects of Gliomas

P. WEINBERG

INTRODUCTION

Technologic advances in the radiologic field, and the greater application of more specialized neuroradiologic procedures have greatly enhanced the accuracy of detection and localization of intracranial lesions. In the investigative work-up of a patient with suspect brain tumor, cerebral angiography is the most frequently employed neuroradiologic diagnostic procedure.

Pneumography, which previously represented the diagnostic study primarily used for localization of tumors, is presently used only if the angiogram and brain scan fail to localize the lesion. The diagnosis of brain tumors by angiography is dependent on the demonstration of displaced vessels and the pathologic vascularization of the lesion. The pathologic vascular patterns of many intracranial neoplasms are often sufficiently characteristic in their appearance to permit a specific histologic diagnosis of the pathologic process. It is the purpose of this paper to review the angiographic descriptions of other investigators together with a presentation of our own case material. This will include a discussion of the various angiographic patterns that may be seen with cerebral gliomas.

ANGIOGRAPHIC TECHNIQUE

The approach that is used in performing cerebral angiography in our Department is determined by the patient's age, clinical status and the type of information we are trying to obtain from this study. Whenever possible, it was elected to perform selective catheterization of the cerebral vessels via the femoral approach. This method is particularly useful in cases that necessitate multiple vessel study and in most cases of suspect brain tumor a bilateral carotid study is obtained. Catheter angiography may be difficult to perform, especially in the elderly patient with extensive atherosclerotic disease or marked tortuosity of the vessels. In these cases, we usually elect to perform percutaneous carotid or retrograde brachial angiography.

The diagnostic accuracy of angiography depends to a certain extent on the technical quality of the examination. If one obtains cerebral angiograms of optimal quality, demonstrating all phases of the cerebral circulation, a high degree of accuracy of diagnosis will be maintained. In most cerebral angiograms, exposures are made at the rate of two films per second for the initial three seconds, followed by one film per second for a period of seven seconds. It has been our experience that this prolonged angiographic series results in better visualization of the deep venous system. Abnormalities in the venous phase of a cerebral angiogram are as important in the diagnosis of a lesion as are abnormalities in the arterial phase. In most cases of supratentorial mass lesions there are changes in both the arterial and venous phases; however, in occasional cases the essential features may be confined to one or the other phase. Since the subependymal veins are the only vessels that course in the walls of the ventricles (1), an evaluation of these veins is most important in assessing ventricular size or in the detection of mass lesions adjacent to the ventricle.

There are special techniques that have been employed to enhance the diagnostic accuracy of cerebral angiography. These include magnification angiography, stereoscopic studies, angiotomography and the subtraction process. Magnification angiography, which results in improved visualization of small intracranial vessels that are 250-300 microns in size, is used most often in the further investigation of an area of suspect abnormal vascularity. Our neuroradiologic facilities do not include special equipment for obtaining stereoscopic examinations. This method has been demonstrated by other investigators to be helpful in some cases of vascular malformations and tumors. Angiotomography has been used in a small number of cases to better demonstrate the origin of an aneurysm and its relationship to adjacent vessels as well as in some cases of tumors both in the supratentorial compartment and posterior fossa. The subtraction technique is a valuable process which is routinely employed in all cerebral angiograms. With this method the overlying bony densities of the skull are eliminated resulting in improved visualization of the vasculature of the brain.

DYNAMICS OF CEREBRAL CIRCULATION

In the interpretation of cerebral angiograms it is essential to evaluate not only the morphology of the cerebral vasculature, but also the dynamics of cerebral circulation as well. According to GREITZ (2) the cerebral circulation time was measured as the time that it takes from maximum concentration of the contrast material in the carotid siphon to maximum concentration of the contrast material in the parietal veins. According to GREITZ the circulation time in normal cases is three to seven seconds with a mean arteriovenous circulation time of 4.13 seconds with a standard deviation of ±0.74 seconds. In order to accurately assess the circulation time one must use a standardized technique which depends upon constancy of all technical factors including injection time, injection pressure, viscosity of the contrast material and the amount and concentration of the contrast material.

TAVERAS and WOOD (3) referred to this method of measurement as the arteriovenous circulation time to differentiate it from the circulation time as measured in the arterial, intermediate (capillary) and venous phases. The arterial phase of a carotid angiogram varies between 1.0 and 2.5 seconds with an average arterial circulation time of 1.5 seconds. With a standardized technique an arterial circulation time greater than 2.5 seconds is usually abnormal. The capillary phase corresponds to the period of time between visualization of the distal branches of the arteries and faint visualization of the origin of the frontal veins. This phase of the angiogram generally lasts between 0.5 and 1.0 second. The venous phase begins with initial visualization of the superficial cerebral veins and continues until all of the contrast material is out of the veins. The frontal veins usually fill prior to the parietal veins, or the frontal and parietal veins may appear simultaneously on the angiogram. Filling of the parietal veins before the frontal veins is often an abnormal finding indicating either a slowing of circulation in the frontal region or an increase in the speed of circulation in the parietal region. The deep venous structures usually appear later than the superficial cerebral veins, however, on occasion the deep cerebral veins may fill at the same time.

ANGIOGRAPHY OF THE POSTERIOR FOSSA

Vertebral angiography has been proven to be a most useful neuroradiologic procedure for the evaluation of posterior fossa abnormalities. With our increasing knowledge of the arterial and venous anatomy of the posterior fossa, we are now able to diagnose many lesions that previously required the performance of pneumoencephalography. We therefore perform posterior fossa angiography as the initial neuroradiologic procedure in the investigation of a suspect abnormality, and if necessary a limited air study may be done as a follow-up examination. The detection of a posterior fossa abnormality often requires excellent visualization of all the vessels of the vertebral-basilar system and for this reason we perform vertebral angiography via the femoral catheter route whenever possible.

The posterior fossa is arbitrarily divided into an anterior and a posterior compartment as established on the lateral view using certain fairly constant anatomical landmarks (4). A line constructed between the precentral cerebellar vein and the choroidal portion of the posterior-inferior cerebellar artery separates the posterior fossa into the two compartments. The contents of the anterior compartment include the brain stem, the subarachnoid cistern anterior to the brain stem and the cerebellopontine angle cistern. The contents comprising the posterior compartment include the cerebellum, fourth ventricle, posterior and lateral subarachnoid spaces and the prevermis spaces both superiorly and inferiorly including the fourth ventricle (4).

There are many important anatomical landmarks in the arterial and venous phases of a vertebral angiogram that must be carefully evaluated. Variations in the course and distribution of the posterior-inferior cerebellar artery are numerous and an appreciation of the many normal configurations of this vessel is necessary in order to recognize any pathologic deviations (5). It is nevertheless possible to establish a typical or standard configuration of the posterior-inferior cerebellar artery and thereby recognize displacements of this vessel. The course of the posterior-inferior cerebellar artery is intimately related to the medulla, the inferior aspect of the fourth ventricle, the tonsils, and the inferior aspect of the vermis and cerebellar hemispheres. Posterior displacement of this vessel indicates the presence of a mass lesion either in the lower brain stem or basal cisterns, whereas anterior displacement indicates the presence of a mass lesion arising from the cerebellum, fourth ventricle or posterior cisterns. The angiographic appearance of the veins in the posterior fossa has been described in great detail by HUANG et al. (6, 7).

One of the most useful veins of the posterior fossa is the precentral cerebellar vein. This venous structure originates in the precentral cerebellar fissure between the central lobule of the vermis and the lingula. The first portion of this vein within the precentral cerebellar fissure lies behind the aqueduct and courses in a parallel fashion to the roof of the fourth ventricle. At the point where this vein leaves the precentral cerebellar fissure it lies behind and below the inferior colliculi at which level it courses superiorly and posteriorly in front of the culmen of the superior vermis. This vein then continues its upward course until it joins the posterior aspect of the great vein of Galen. After identifying the precentral cerebellar vein, a line is drawn between the tuberculum sella and the torcula (Twining's line). A second line is then drawn tangentially through the bend of the precentral cerebellar vein (level of inferior colliculi) and perpendicular to Twining's line. This perpendicular line should fall at the midpoint of Twining's line with a standard deviation of ±5% of its length (8). Tumors that are located in the brain stem or originate from the basal subarachnoid cisterns (anterior compartment) may produce a posterior displacement of the precentral cerebellar vein, whereas tumors originating in the vermis or cerebellar hemispheres (posterior compartment) tend to produce anterior displacement of this vein.

The posterior-inferior cerebellar artery and precentral cerebellar vein represent only two of many important anatomical landmarks that are demonstrated with vertebral angiography. However, it is not possible within the scope of this presentation to further elaborate on these findings.

RADIOLOGIC SIGNS OF GLIOMAS

In adults, approximately 70% of all intracranial tumors are supratentorial, whereas in children, approximately 70% of all intracranial tumors are infratentorial (9). This presentation of case material is limited to a discussion of the major types of gliomas. Approximately 50% of all intracranial tumors are classified in the glioma group. Our series of case material corresponds closely with the statistics as presented by ZIMMERMAN (10). During the period of the past nine years, a total of 258 gliomas of various types have been seen at our institution. This figure represents approximately 40% of intracranial tumors of all types in our series. Since the patient population in our hospital is almost exclusively an adult one, our statistics are probably not representative of the general population.

SKULL FILM EXAMINATION

Any patient who is suspected of having a brain tumor must have a plain film examination of the skull as the initial radiologic examination in the investigative work-up. Calcification is seen in approximately 10-15% of all tumors and may occur in the form of small punctate densities or coarse linear calcification. Oligodendrogliomas demonstrate the highest incidence of calcification occuring in approximately 50% of the cases, followed by astrocytomas with a 20-25% incidence of calcification. Although astrocytomas do not calcify as frequently as oligodendrogliomas, astrocytomas are nevertheless the leading cause of calcification in the glioma group since this tumor is the most common type of tumor in this group (3). Calcification is seen infrequently with ependymomas and when demonstrated is more often in association with the ependymomas arising in the lateral ventricles or adjacent white matter. Another localized finding of glioma as seen on the plain skull series may be pressure erosion of the inner table of the skull which occurs as an infrequent finding with an occasional intracerebral glioma.

In the presence of increased intracranial pressure, the radiologic findings will vary with the age of the patient. In the adult there may be demineralization of the floor of the sella turcica and of the dorsum sellae. In children the first sign of increased intracranial pressure is usually a separation of the sutures which may occur up to the age of approximately fifteen years with the finding of a demineralized sella occurring at a later stage. The pineal body is calcified in about 50-60% of adults and displacement of this calcification from one side to the other usually indicates the presence of a mass lesion.

BRAIN SCAN

In the diagnosis of intracranial tumor it is important to establish the exact location and degree of extension of the lesion, and whenever possible, to determine the histologic nature of the tumor. In most cases a brain scan is obtained prior to performing any contrast studies. In some cases the results of the brain scan may obviate the necessity for performing neuroradiologic examination; an example is the case of a patient with a suspect metastatic disease with multiple lesions demonstrated on the scan. We always interpret the angiographic findings together with the results of the brain scan and in our experience this combined evaluation has resulted in an increased accuracy both in detection of lesions as well as the determination of histologic type.

PNEUMOENCEPHALOGRAPHY

A comparison of the relative merits of angiography and air studies in diagnosing lesions in the supratentorial region revealed that pneumography might be very helpful in demonstrating lesions located in the periventricular region as well as within the ventricular system. The majority of intraventricular tumors belong to the glioma group and are usually astrocytomas or oligodendrogliomas. Oligodendrogliomas commonly have an irregular surface and may extend through the septum pellucidum presenting as biventricular masses. Astrocytomas do not present as characteristic an appearance as oligodendrogliomas and often arise from the basal ganglia or other adjacent structures and extend into the ventricular system.

Tumors originating in the anterior portion of the third ventricle are frequently indistinguishable from tumors originating in the hypothalamic region. Astrocytomas and ependymomas are among the tumors that are encountered in this region and with increasing size they might obliterate the inferior portion of the third ventricle. If a mass lesion in the hypothalamic region presents an irregular superior surface as outlined by air in the third ventricle this is usually indicative of an infiltrating glioma.

Tumors originating within the fourth ventricle such as ependymomas frequently invade adjacent brain structures, and medulloblastomas and astrocytomas arising in the periventricular structures often invade the fourth ventricle. It may therefore be very difficult to determine the site of origin of these tumors. When ependymomas are confined to the fourth ventricular cavity they have a fairly typical appearance with air outlining a mass lesion arising from the ventricular floor with associated marked dilatation. More often the lumen of the fourth ventricle is occluded in which case there is no characteristic pneumoencephalographic appearance. In performing pneumoencephalograms for a suspect posterior fossa mass we routinely obtain lateral and anterior-posterior tomograms in the erect position.

Primary glial tumors of the optic nerves or optic chiasm occur primarily in children. Plain film examination of the optic foramina may demonstrate enlargement of the optic canal. These tumors do not have any characteristic pneumoencephalographic features that distinguish them from other suprasellar mass lesions. To best demonstrate a lesion in the suprasellar region we routinely obtain lateral and antero-posterior tomograms in the brow-up position. Optic gliomas usually obliterate the prechiasmatic region and may produce splaying or complete obliteration of the anterior third ventricle.

The most significant finding in the pneumographic evaluation of cerebellar hemispheric and vermis gliomas is the pathologic kink of the aqueduct. In lesions of the vermis, the aqueduct remains in the midline, whereas, this structure is displaced laterally when the glioma is located in one of the cerebellar hemispheres.

ANGIOGRAPHY PATTERNS OF GLIOMAS

The possibility of demonstrating a tumor and establishing a differential diagnosis by cerebral angiography is dependent on the fact that many tumors have a vascular pattern that differs from the vascularity of normal brain tissue. The demonstration of vessel displacements only indicates the presence of a space occupying lesion, whereas the abnormal vascular pattern may indicate the histology of the tumor. The vessels supplying a tumor may represent existing arteries that have become enlarged as a result of increased blood flow through the tumor, or possibly represent new vessel formations. There may also be enlargement of pre-existing veins or the development of new venous channels draining the tumor site. These vascular changes may result in tortuous, dilated vessels having an irregular lumen, a "beaded" vessel appearance, sinusoidal formation and arteriovenous anastomosis. The tumor blush that may be seen with gliomas is probably the result of contrast filling of dilated cortical arteries and veins (11).

We are often confronted with the dilemma of how much importance to attach to a borderline type of change, such as a questionably stretched vessel or a suspect localized abnormal circulation time (12). It is possible to have a normal appearing angiogram in the presence of an intracranial tumor, particularly in the case of an intraventricular lesion or an infiltrating glioma (13).

In many tumor cases the observation of local slowing or increased speed of circulation time has helped establish the diagnosis of a tumor (3). With respect to the local arterial circulation time, the presence of an area of delayed appearance of one or more arteries occurs primarily in intra-axial tumors such as gliomas and metastasis and rarely in extra-axial tumors such as meningiomas. A delay in the appearance of local venous filling without evidence of tumor circulation may be seen with intra-axial tumors such as gliomas or metastasis as well as with meningiomas. A delayed appearance of local veins in association with tumor vascularity occurs more frequently in meningiomas than in gliomas.

The presence of an increased speed of local circulation in the presence of tumor vascularity usually indicates that the lesion is a malignant one, such as glioblastoma multiforme or metastasis. Another abnormality in the circulation time is the presence of an early filling vein without evidence of tumor vascularity. This finding has been demonstrated in various types of tumors as well as in other lesions including infarcts, abscesses and hematomas. A regional abnormality in the circulation time may be the only finding that is seen with cerebral angiography and therefore, demonstrates the importance of carefully evaluating the venous filling pattern.

Astrocytomas

Astrocytomas account for about 25-30% of all cerebral gliomas; and in the adult, approximately 8% of all gliomas occur in the cerebellum, whereas in children, cerebellar gliomas comprise about 30% of all gliomas. These tumors are usually slow growing and present as avascular mass lesions. The circulation time, if abnormal, is usually prolonged through the tumor site. Cerebral hemispheric astrocytomas are usually diffuse tumors that may occur anywhere within the cerebral hemisphere, although the occipital lobes are relatively spared (14). Astrocytomas occurring in the cerebellar hemispheres occur predominantly in the first two decades of life; they are often circumscribed with a tendency to be grossly cystic and usually remain histologically benign. It is also not uncommon for these tumors to be solid and diffusely infiltrating (14, 15). Regardless of the site of occurrence, these astrocytomas are generally avascular mass lesions and are diagnosed by vessel displacements and possible localized slowing of arterial and venous filling.

Case Presentation. The patient was a 59-year-old, right handed, white female who was admitted to the hospital with the presenting complaint of mental confusion first noted four months prior to admission. The patient also complained of dull, generalized headaches ten days prior to admission, with no other associated complaints. Examination revealed that the patient was disoriented as to time and place and had an inappropriate affect. There was no evidence of any motor weakness and sensation was intact. There was blurring of both disc margins and the patient had bilateral snout and bilateral Babinski signs. Skull examination was normal. A brain scan demonstrated the presence of a well circumscribed area of greatly increased activity in the left frontal region which extended at least to the midline. Left carotid angiography (Fig. 1) demonstrated the presence of a large avascular mass lesion in the frontal lobe producing vessel displacements with no evidence of tumor vascularity, tumor stain or arteriovenous shunting. There was delayed filling of the superficial veins in the frontal region secondary to the mass lesion. The angiographic diagnosis was astrocytoma of the frontal lobe. At surgery, a large solid left frontal tumor was found and the histologic diagnosis was that of astrocytoma.

Glioblastoma Multiforme

Glioblastoma multiforme represents the most anaplastic form of glial cell tumor, and it has been classified as astrocytoma grades III and IV by KERNAHAN and SAYRE (16). This tumor accounts for approximately 50% of all primary gliomas, has a peak incidence between forty-five and fifty-five years of age and tends to affect males more frequently than females by a ratio of 3:2 (9). Our series of cases demonstrated that the most common site of occurrence is in the frontal lobe with the temporal lobe representing the second most frequent site of involvement.

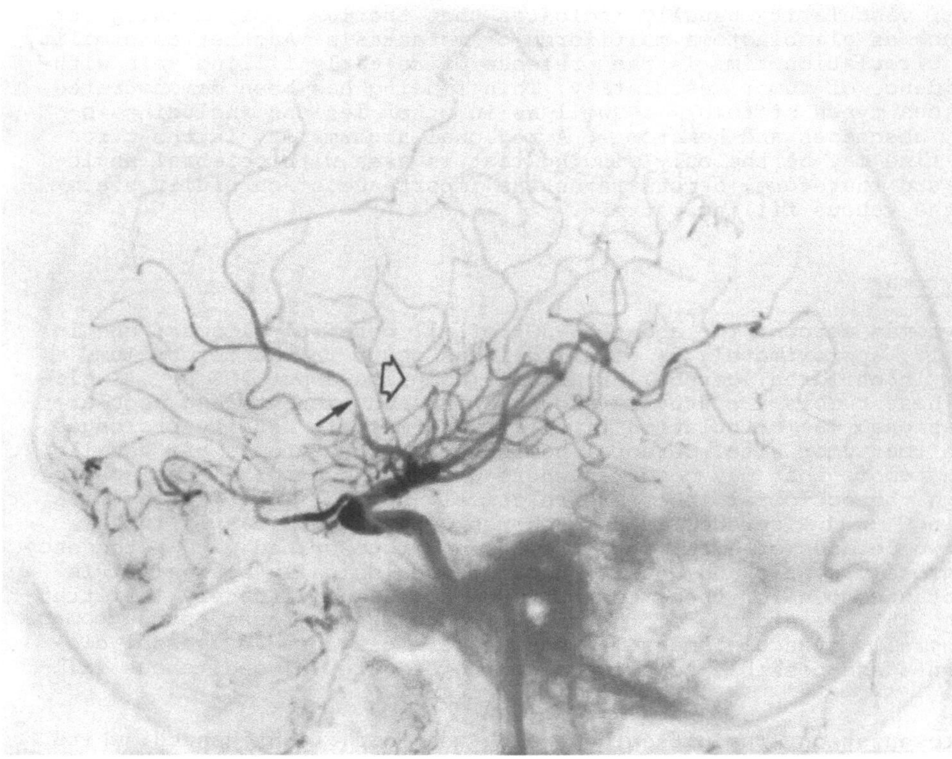

Fig. 1. Carotid angiogram, lateral projection. This film in the early arterial phase demonstrates posterior displacement of the middle cerebral vessels within the anterior aspect of the Sylvian fissure (open arrow) and posterior displacement of the proximal anterior cerebral artery with a localized site of narrowing of this vessel (closed arrow). There is also stretching of orbital frontal and frontal polar branches of the anterior cerebral artery and posterior and inferior displacement of the supraclinoid segment of the internal carotid artery. There is no evidence of tumor vascularity

The deeper structures including the basal ganglia and the corpus callosum are often invaded; with tumors occurring in the deep frontal region there may be extension to the contralateral hemisphere through the corpus callosum presenting a "butterfly" pattern. The cerebellum and brain stem represent infrequent sites of involvement of this tumor.

WICKBOM (17) described different vascular patterns that may be demonstrated in the presence of a glioblastoma and presented an arteriographic classification as outlined in Table 1.

WICKBOM (17) also described another vascular abnormality characterized by the presence of a non-vascular area which is surrounded by a peripheral highly vascularized zone. He attributed this angiographic vascular pattern to the development of necrotic and cystic changes that are often present in glioblastomas. Although this ring-like vascular pattern is seen most frequently in cases of glioblastoma multiforme, it is not in itself a pathognomonic finding. The same ring-like vascular pattern has been described with other lesions including abscesses and meningiomas (17, 18), metastatic carcinoma, subdural abscess, malignant astrocytomas and cerebral hemangioblastoma (19).

Table 1

Glioblastomas: Wickbom (10)

Type I	Irregular coursing vessels of varying size and arteriovenous fistulas
Type II	Irregular tumor vessels similar to those in Type I but without arteriovenous fistulas
Type III	Tumor vessels that are shorter and smaller than Type I and II with small aneurysmal dilatations
Type IV	Short, thin vessels with slightly irregular lumens

The diffuse homogeneous or nearly homogeneous blushing that is seen in some cases of glioblastoma multiforme has been attributed to altered gray matter vascularity with vasodilatation as well as neovascularity from gray matter arteries and veins (11). Other lesions such as metastases and cerebral infarcts can also result in similar gray matter involvement with associated homogeneous staining. LASSEN (20) has described a "luxury perfusion" seen in angiograms where an area of brain infarction is outlined by capillary blush. It is postulated that an elevated pCo_2 occurring in an area of ischemic brain acts as an acidotic stimulus which results in vasocapillary dilatation. The hypervascularity in gray matter that is seen in association with tumor infiltration is not unlike that observed in brain infarcts (21).

At times one can obtain greater knowledge regarding the pathophysiology of circulation in tumor cases by performing functional tests during angiography (22). With respect to alterations in the pCo_2 it was demonstrated that angiography performed during active hyperventilation with a reduction of the pCo_2 resulted in improved visualization of A-V shunting and tumor stains in vascularized tumors. In cases of malignant brain tumors with an extensive network of pathologic vessels and A-V shunts, functional tests provided no additional information concerning the tumor since the diagnosis was already clear on the standard angiographic series. In those cases of poorly vascularized tumors, or tumors with questionable A-V shunting, angiography obtained during hyperventilation may result in a denser tumor blush and a more clearly demonstrable early-draining vein. In our case material, we have made frequent use of the hyperventilation study together with magnification angiography, especially in cases of questionable arteriovenous shunting, suspect tumor vessels or uncertain tumor stain.

Case Presentation. The patient was a 55-year-old, right handed, white female who complained of a two week history of severe headaches and was noted to be confused and forgetful during the week prior to hospital admission. There was no history of seizures or complaint of weakness. Neurologic examination revealed blurring of the left disc, slight weakness of the right upper and lower extremities and a right Babinski. The patient was also confused and disoriented and displayed poor memory and inappropriate behavior. A brain scan, obtained at another hospital, was reported as demonstrating an increased radioactive uptake in the left frontal region. Bilateral carotid angiography was performed. The left carotid angiogram (Fig. 2) demonstrated the presence of a large mass lesion situated within the frontal polar region and inferior frontal gyrus producing marked displacement of vessels. There was an extensive network of irregular, dilated, tortuous vessels and "beaded" appearing vessels predominantly situated within the periphery of the mass lesion with multiple sites of arteriovenous shunting both into the superficial and deep draining veins. The angiographic diagnosis was frontal lobe glioblastoma extending into the anterior basal ganglion region. Surgery was performed and revealed the presence of a very malignant-appearing tumor in the frontal lobe extending deep into the basal ganglion region, and the histologic diagnosis was glioblastoma multiforme.

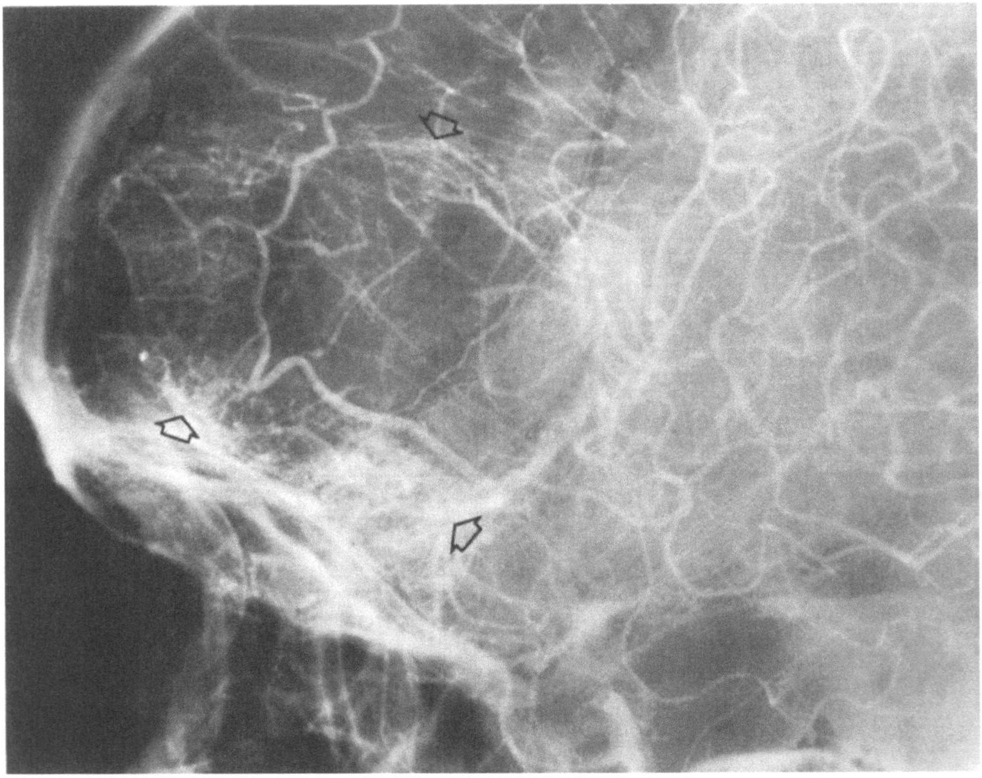

Fig. 2. Carotid angiogram, lateral projection. This film obtained from the mid-arterial phase of a magnification angiogram demonstrates the presence of a mass lesion in the frontal lobe region with extensive neovascularity in the periphery of the lesion (open arrows)

Oligodendrogliomas

Oligodendrogliomas were originally described by BAILEY and BUCY (23) and constitute approximately 5-7% of all primary intracranial gliomas. In a review of the angiographic findings in fifty-three patients with oligodendrogliomas (24), it was noted that because of the similar pathologic blood vessels encountered in this tumor and in astrocytomas that a differential diagnosis was not possible. These same investigators reported the presence of pathologic vascularity in twenty-one of the fifty-three cases with the abnormal vascularity characterized by the presence of "beaded" vessels. This pathologic vascularity was most often demonstrated in the arterial and capillary phases, and the presence of arteriovenous communications was noted in approximately 10% of the cases. These tumors usually arise in the cerebral hemispheres where they are most often encountered in the frontal lobe (45%) with other common sites of occurrence including the temporal lobe (36%) and the parietal lobe (17%). The most uncommon site of involvement is the occipital lobe (2%) (25). Some oligodendrogliomas that are situated near the surface of the brain may invade the overlying leptomeninges and therefore mimic the picture of an endotheliomatous meningioma.

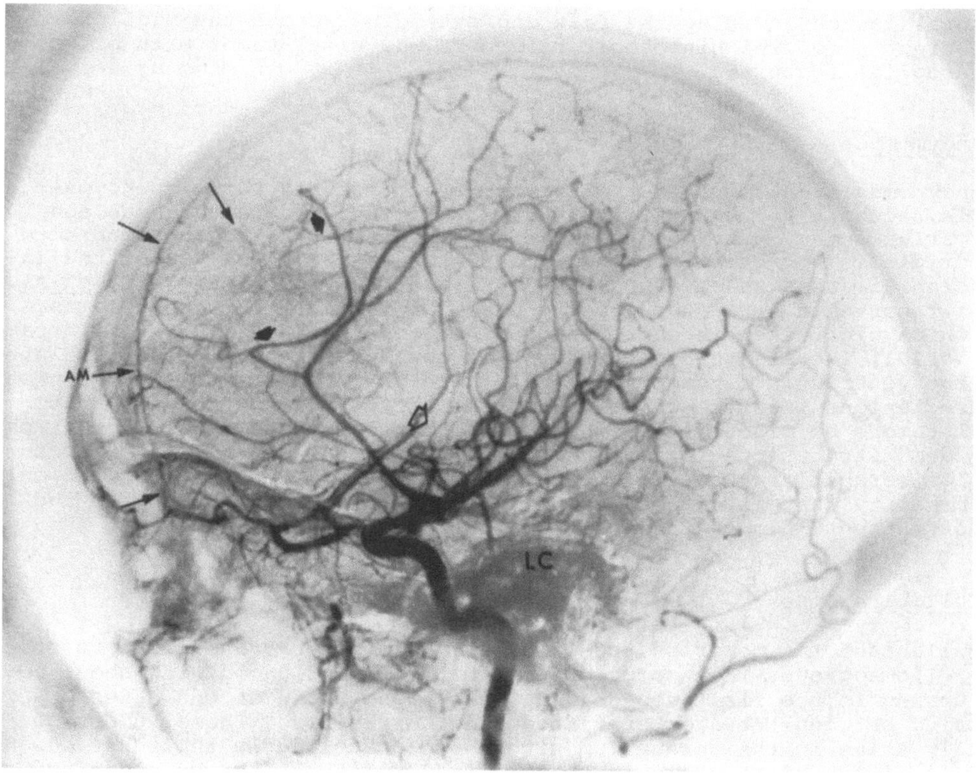

Fig. 3. Carotid angiogram, lateral projection. This film from the arterial phase demonstrates the presence of a large frontal lobe mass lesion with posterior displacement of the middle cerebral branches within the anterior aspect of the Sylvian fissure (open arrow), posterior displacement and stretching of anterior cerebral artery branches (closed arrowheads) and an enlarged anterior meningeal branch arising from the ophthalmic artery which extends superiorly to the tumor site (closed arrows)

Case Presentation. The patient was a 55-year-old, right handed, white female who presented with a two week history of inappropriate behavior, characterized by saying strange things and displaying a loss of affect. The patient complained of recent onset of left sided headaches and decreased visual acuity in the left eye. Neurologic examination revealed the presence of a right Babinski sign with no evidence of motor weakness. The patient displayed an inappropriate affect, confusion and poor recent memory. A brain scan demonstrated a large area of increased radioactive uptake in the midline of the frontal region. Bilateral carotid angiography was performed (Fig. 3) and demonstrated a large mass lesion in the frontal polar region predominantly on the left side and extending across the midline. There was a large anterior meningeal artery branch arising from the left ophthalmic artery that contributed to the blood supply of this mass lesion. There were also "beaded" tumor vessels arising from branches of both the right and left anterior cerebral arteries. There was delayed filling of the superficial frontal veins secondary to the mass lesion. The primary angiographic diagnosis was that of malignant astrocytoma with invasion of the falx with a secondary diagnosis of meningioma. Surgery revealed a large left frontal lesion invading the falx and extending across the midline to the right side. The histologic diagnosis was glial tumor with oligodendroglial features.

Ependymomas

Ependymomas are tumors that originate from differentiated ependymal cells, account for approximately 6% of all intracranial gliomas and may arise in any part of the ventricular system. The most common site of origin is in the fourth ventricle which accounts for 70% of the intracranial cases as reviewed by FOLKES and EARL (25). KRICHEFF, et al. (26) reported a series of seventy cases of intracranial ependymomas, of which eighteen were supratentorial in location and fifty-two infratentorial. The eighteen supratentorial ependymomas were grossly related to the ventricular system; however, there was a major growth of ten of these lesions into the white matter. All fifty-two of the infratentorial tumors originated from the fourth ventricle. Vertebral angiography may demonstrate anterior or inferior displacement of the posterior inferior cerebellar artery. Ventricular dilatation may represent the only abnormality that is demonstrated with carotid angiography unless the tumor extends into the adjacent brain tissue.

Medulloblastomas

Medulloblastomas comprise approximately 6% of intracranial tumors of the glioma group. They occur predominantly in childhood with about 50% occurring in the first decade and another one-third of the cases occurring between the ages of fifteen and thirty-five. These tumors generally arise in the cerebellum with those occurring in the first decade originating predominantly in the midline, and those in the older age group more likely occurring in the cerebellar hemispheres. The tumors that arise from the vermis usually originate from the region of the posterior medullary velum in the nodulus of the vermis and have a tendency to invade the fourth ventricle and the cisterna magna region. A knowledge of the anatomy of the posterior fossa circulation, particularly with reference to the posterior inferior cerebellar artery may allow one to diagnose the presence of a mass lesion in the region of the fourth ventricle even in the absence of tumor vascularity.

CONCLUSION

Neuroradiologic procedures are highly accurate in the detection and localization of intracranial lesions. Nevertheless, despite the extreme refinement of these procedures, there are still lesions that elude our detection. The correlation of neurodiagnostic and neuro-nuclear studies has increased our ability to localize these elusive lesions. With the advent of sectional tomography of the brain as demonstrated with the use of the EMI Scanner, we can look forward to an even greater improvement in the neuroradiologic diagnosis of brain tumors.

REFERENCES

1. WOLF, B. S., HUANG, Y. P.: Diagnostic value of cerebral veins in mass lesions of the brain. The Radiologic Clinics of North America Vol. IV, 1, 117-130. Philadelphia and London: W. B. Saunders Co. 1966.
2. GREITZ, T.: A radiologic study of the brain circulation by rapid serial angiography of the carotid artery. Acta Radiol., Suppl. 140 (1956).
3. TAVERAS, J. M., WOOD, E. H.: Diagnostic neuroradiology. Baltimore: Williams and Wilkins 1964.
4. DAVIS, D. O., ROBERSON, G. H.: Angiographic diagnosis of posterior fossa mass lesions. Seminars in Roentgenology 6, 1 (1971).
5. WOLF, B. S., NEWMAN, C. M., KHILNANI, M. T.: The posterior inferior cerebellar artery on vertebral angiography. Am. J. Roentgenol. Radium Therapy Nucl. Med. 87, 322 (1962).
6. HUANG, Y. P., WOLF, B. S.: Veins of posterior fossa-superior or galenic draining group. Am. J. Roentgenol. Radium Therapy Nucl. Med. 95, 808 (1965).
7. HUANG, Y. P., WOLF, B. S., ANTIN, S. P., OKUDERAS, T.: Veins of posterior fossa-anterior or petrosal draining group. Am. J. Roentgenol. Radium Therapy Nucl. Med. 104, 36 (1968).
8. HUANG, Y. P., WOLF, B. S.: Precentral cerebellar vein in angiography. Acta Radiol. 5, 250 (1966).
9. RUBINSTEIN, L. J.: Tumors of the central nervous system. A. F. I. P. (1972).
10. ZIMMERMAN, H. M.: The natural history of intracranial neoplasm, with special reference to the gliomas. Am. J. Surg. 93, 913 (1957).
11. SCATLIFF, J. H., RADCLIFFE, W. B., PITTMAN, H. H., PARK, C. H.: Vascular structure of glioblastomas. Am. J. Roentgenol. Radium Therapy Nucl. Med. 105, 795 (1969).
12. DI CHIRO, G.: Relative value of air studies, angiography and radioisotope scanning in the diagnosis of glial intracranial tumors. Progress in Neurological Surgery, 2, 292-317. Chicago: Karger, Basel and Yearbook 1968.
13. PETERSON, H. O.: Reliability and limitations of x-ray diagnosis of intracranial neoplasms. Am. J. Surg. 93, 941 (1957).
14. RUSSELL, D. S., RUBINSTEIN, L. J.: Pathology of tumors of the nervous system. Baltimore: William and Wilkins 1971.
15. BUCY, P. C., GUSTAFSON, W. A.: Structure, nature and classification of cerebellar astrocytomas. Am. J. Cancer 35, 327 (1939).
16. KERNAHAN, J. W., SAYRE, G. P.: Tumors of the central nervous system. FASCICLE 35, Atlas of Tumor Pathology. Washington: Armed Forces Institute of Pathology 1952.
17. WICKBOM, I.: Angiographic determination of tumor pathology. Acta Radiol. 40, 529 (1953).
18. LINDGREN, E.: Radiologic examination of brain and spinal cord. Acta Radiol., Suppl. 151 (1957).
19. FEIRING, E. H., SHAPIRO, J. H., LUBETSKY, H. W.: The ring-like vascular pattern in cerebral arteriography. Radiology 89, 385 (1963).

20. LASSEN, N. A.: Luxury-perfusion syndrome and its possible relation to acute metabolic acidosis localized within the brain. Lancet 1966 II, 1113.
21. ROBINSON, F., PORRO, R. S., SCATLIFF, J. H.: Angiographic recognition of occipital lobe infarction. Neurology 16, 1016 (1966).
22. HUBER, P.: Functional tests in angiography of brain tumors. Neuroradiology 1, 132 (1970).
23. BAILEY, P., BUCY, P. C.: Oligodendrogliomas of the brain. J. Path. Bacteriol. 32, 735 (1929).
24. SPETTOWA, S., KUGMIDERSKI, J.: Angiography of oligodendrogliomas. Acta Med. Polona IX, 4, 481 (1968).
25. FOKES, E. C. JR., EARL, K. M.: Ependymomas: clinical and pathological aspects. J. Neurosurg. 30, 585 (1969).
26. KRICHEFF, I. J., BECKER, M., SCHNECK, S. A., TAVERAS, J. M.: Intracranial ependymomas. Am. J. Roentgenol. Radium Therapy Nucl. Med. 91, 167 (1964).

Computerized Tomography in Relation to the Diagnosis of Gliomas

M. S. HUCKMAN

Computerized axial transverse tomography (CT) is rapidly becoming an important and reliable tool for the radiographic detection of gliomas as well as other intracranial masses. The concept was developed by Mr. G. N. HOUNSFIELD (3) and has provided radiologists with a virtually non-invasive method of producing a scan of the variations in x-ray absorption in multiple sections of the brain. The system described by HOUNSFIELD consists of an x-ray tube with a collimated beam mounted on a frame directly opposite two collimated sodium iodide crystal detectors. The patient's head is placed in a rubber cap surrounded by water which lies between the x-ray tube and the detector crystals. The x-ray beam and the detectors scan the patient's head in a linear fashion taking 160 readings of x-ray transmission through the head. After each scan, the frame containing the x-ray tube and detector is rotated 1° and the process is then repeated. After the frame has rotated 180° and taken 160 readings for each degree of rotation, there will be a total of 28,800 readings of x-ray transmission. These readings are then processed by a mini-computer and a picture is displayed on a cathode ray tube in the form of an 80 x 80 matrix in which each point has a numerical value indicating the coefficient of absorption for a volume of material in the slice measuring 3mm x 3mm x 13mm. The absorption coefficients may be printed out on a line printer or viewed on a cathode ray tube. By varying the window width and window level on the cathode ray tube, it is possible to make accurate measurements of the coefficient of absorption of any point in the matrix.

The scale used on the print-out is arbitrary and is related to water which is arbitrarily assigned the numerical absorption value (or NAV) of 0 and air is given a NAV of -500 units. Bone calcification is approximately +500 units. According to HOUNSFIELD (3), white matter has a NAV of 12, gray matter is 18 and congealed blood is 20 to 30. Cerebrospinal fluid is slightly greater than zero probably due to its mineral content.

MATERIAL

The cases to be considered in this report are the known gliomas which occurred in the first 1100 CT scans performed at the Rush-Presbyterian-St. Luke's Medical Center. These consisted of 13 primary gliomas and 1 metastatic cerebral medulloblastoma which appeared fourteen years after surgery and radiation for a cerebellar medulloblastoma. Thirteen of the cases were surgically proved. The remaining case was felt to be a glioma on the CT scan and arteriography performed at another institution was reported as "an infiltrating deep glioma". This patient refused surgery and was given radiation therapy which resulted in a considerable improvement of her symptoms. The pathologic distribution of our cases is shown in Table 1.

Table 1. Case material

Pathologic diagnosis	No. of cases
A. Oligodendroglioma	3
B. Grade II Astrocytoma	3
C. Grade III Astrocytoma	1
D. Glioblastoma Multiforme	4
E. Medulloblastoma	2
F. Unproved	1

Due to the fact that this technique of scanning was quite new, there were no available criteria with which we could attempt histologic grading. Therefore, much of our initial experience was empiric and rather than try to grade abnormal masses histologically, we merely tried to classify them as to size, location, whether or not they occupied space, whether they contained calcification, blood or edema and whether the numerical absorption value suggested a cyst (a NAV 1-10), infarct (NAV 5-12), glioma or metastases (NAV 11-15), meningioma (NAV 20-35) or hemorrhage (NAV 30-45).

Table 2. Computerized tomography reports

Pathologic diagnosis	C. T. diagnosis
1. Oligodendroglioma	Hemorrhage
2. Oligodendroglioma	Glioma or Metastasis
3. Oligodendroglioma	Meningioma
4. Grade II Astrocytoma	Glioma or Metastasis
5. Grade III Astrocytoma	Glioma or Metastasis
6. Grade III Astrocytoma	Glioma or Metastasis
7. Grade III Astrocytoma	Glioma or Metastasis
8. Glioblastoma Multiforme	Glioma or Metastasis
9. Glioblastoma Multiforme	Glioma or Metastasis
10. Glioblastoma Multiforme	Glioma or Metastasis
11. Glioblastoma Multiforme	Glioma or Metastasis
12. Medulloblastoma	Meningioma
13. Medulloblastoma	Negative Scan
14. Unproved	Glioma or Metastasis

Trying to place all of our space-occupying lesions into one of the above categories when they were initially read gave us the following diagnoses listed in Table 2. As you can see fron this table, thirteen cases were recognized on the CT scan as space-occupying lesions. If we were to consider the fact that a diagnosis on the CT scan of glioma or metastases constituted a positive diagnosis, then this was arrived at in ten out of the fourteen cases. One of the medulloblastomas presented as a false-negative CT scan. However, this particular patient had undergone pneumoencephalography two days prior to the CT scan. As a result, there was air in the ventricular system at the time of CT scan and this caused a computer overswing artifact which made the scan virtually unreadable in the area where the lesion would have been expected to appear. The oligodendroglioma which was called a hemorrhage actually had a large hemorrhagic component in the tumor at surgery. The oligodendroglioma which was called a meningioma was markedly calcified and made it difficult to determine the exact density of the underlying soft tissue within the mass. The medulloblastoma which was diagnosed as a meningioma was simply a mistaken assumption that a high density lesion would not be in the group of glioma or metastases.

DISCUSSION

Due to the fact that computerized tomography has been available for only a short period of time, there is no existing extensive study of its use in the diagnosis of gliomas. AMBROSE (1) states that no distinct pattern has emerged from his study of the numerical absorption values in low density lesions. However, he stated that it should be possible to differentiate tumors from non-neoplastic lesions by the presence of necrosis, cyst formation and displacement of normal structures such as ventricles and the pineal gland. He further states that tumors should be distinguishable from infarcts by the fact that tumors are less likely than infarcts to involve white matter. Gliomas should be distinguishable from metastases according to AMBROSE by the absence of multiplicity although solitary metastasis may be difficult to tell from gliomas.

BAKER et al. (2) in their analysis of the first 529 cases examined by computerized tomography at the Mayo Clinic reported 45 gliomas, 50 other primary tumors and 15 tumors of questionable type. In their 529 cases, there were eighteen cases found to be diagnostic errors. One of their false-negative scans turned out to be a parasagittal meningioma. Two of their gliomas presented negative scans. One case which they diagnosed as a temporal meningioma turned out to be a thalamic glioma and one case which was diagnosed as a rostral corpus callosum glioma by computerized tomography was later pathologically identified as a bi-frontal meningioma with edema. In general, their diagnostic error of 3 1/2% was felt to be favorably comparable to the accuracy obtained with the other standard neuroradiologic procedures employed in their institution.

NEW et al. (4) in their first 300 cases encountered 26 gliomas, although their clinically proved cases were all glioblastomas. They reported wide variation in CT patterns not only from case to case but also within different parts of the same tumor. According to them, the more densely packed tissues in glioblastomas showed numerical absorption values the same or higher than normal white matter and if there was a high fluid content, these values were often considerably lower than white matter. They observed tumor margins to be difficult to recognize in many cases due to peripheral edema. Areas of increased numerical absorption value in gliomas were felt to represent sites of hemorrhage. Necrosis and cysts were noted to have lower numerical absorption values than the surrounding gliomatous tissue. They did not detect any calcifications in any of their tumors although they speculated that microscopic calcification should be visible even though it was not seen on the conventional radiograph. They also suggested that some enhancement might be obtained by the administration, prior to scanning, of iodine-containing intravenous contrast media.

With the pathologic diagnoses known, our cases were reviewed by means of localizing areas on polaroid photographs and returning to the number print-out and examining these areas in further detail. This was done in hopes that some pattern might emerge to enable us to more accurately predict the histology of these tumors in the future.

Analyzing the known pathologic entities separately, first let us consider the oligodendrogliomas seen in Fig. 1 as cases 86, 288 and 649.

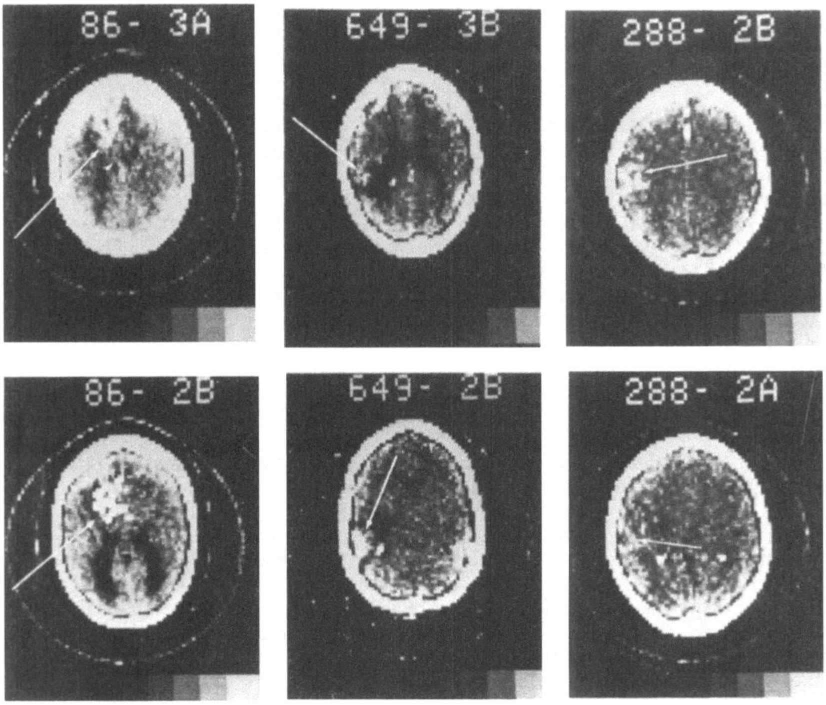

Fig. 1. Case 86 shows a left frontal oligodendroglioma with calcification. Case 288 shows an oligodendroglioma with a large clot surrounding and within the tumor. Case 649 showed a normal plain film although pathologic examination revealed microscopic calcification within this oligodendroglioma

In case 86, the patient had a calcified mass present on the plain film examination for the past five years. The CT scan demonstrated that the calcification had a numerical absorption value of 74.9 and that the soft tissue component of the tumor measured 16.4. We did not initially appreciate the area between the calcification and edema seen on the 3A cut. This area measured approximately 16.4 in average density and might have suggested the diagnosis of a glioma with calcification. We tended to see it as two separate densities, one of the calcification and one of the edema and, therefore, misdiagnosed this as a meningioma. Case 288 presented with a density of 27.5 in the left parietal lobe surrounded by a relatively less dense area . There were several squares of considerably higher density, namely 31 and 32, and this was felt to represent an intracerebral hemorrhage. At surgery, a large blood clot was seen surrounding and within a friable, fleshy vascular tumor. The third case, No. 649, presented with an area of dense calcification measuring 68.5 as seen on the 2B cut. This appears to be surrounded by edema on the 3B cut which measured 12.7 in NAV. While oligodendrogliomas frequently calcify and, therefore, make it difficult to give a histologic diagnosis, in the first and third of these cases where the lesion was obscured by calcification, more careful examination would have revealed that between the calcification and the area of edema, lay areas of relatively diminished density. An area with NAV of 14 on the 2B scan of case 649 and an area with NAV of 16.4 just medial to the calcification seen in the 3A cut of scan 86. These soft tissue values, as will be seen later, are slightly higher than those we encountered in the higher grade astrocytomas.

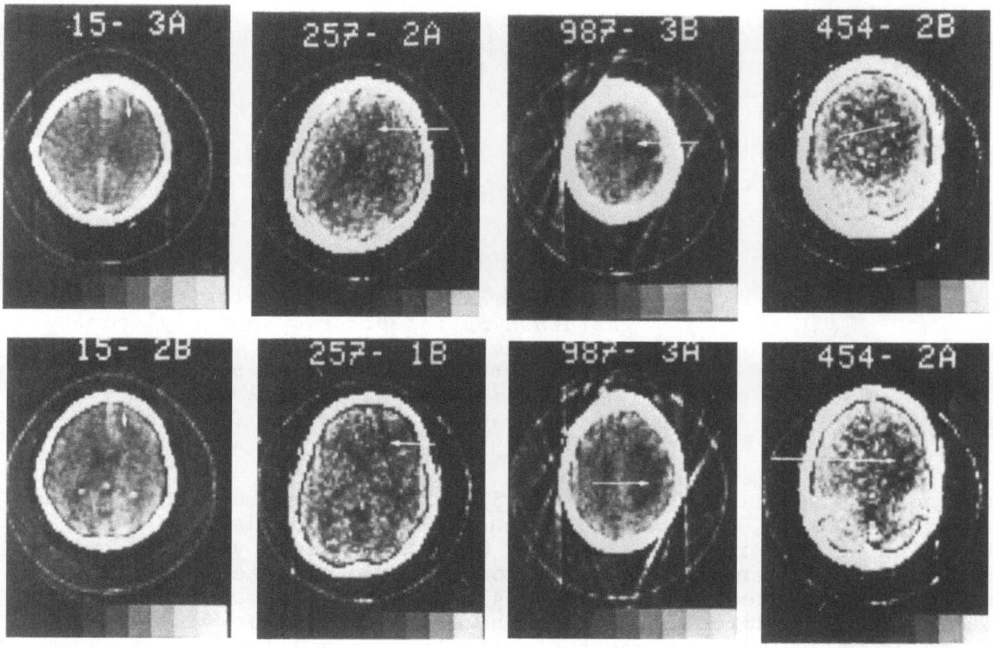

Fig. 2. Case 15 was a Grade III astrocytoma in the right frontal lobe. Case 257 was a Grade III astrocytoma in the frontal lobe. Case 987 was a right parietal cystic Grade III astrocytoma. Case 454 was a Grade II astrocytoma in the right frontal lobe

The Grade II and Grade III astrocytomas appeared quite homogeneous as seen in Figure 2. Case 454 was a Grade II astrocytoma and had a NAV on the average of 13.8. Case 987 had a NAV of 9.9 and at operation, the lesion was reported as being almost entirely cystic containing a viscid fluid. Case 257 was a frontal Grade III astrocytoma, had a numerical absorption value of 13.8 and was described by the surgeon as looking like normal brain with slightly increased consistency and a slightly grayish discoloration. Case No. 15, a right frontal grade III astrocytoma, had an average NAV of 11.7 and was described at surgery as being a yellowish nodule of firm tissue felt clinically to be a meningioma. As you can see by its number, case 15 was one of our earliest cases. Our confidence in diagnosing gliomas was not very great and we were uncertain as to whether what we were seeing truly represented an abnormality in the right frontal lobe. Therefore, the scan was repeated after intravenous injection of 30cc methylglucamine iothalamate. Scan No. 30 (Fig. 3) shows the enhancement obtained in the image by just a small amount of iodine-containing material injected intravenously prior to scanning.

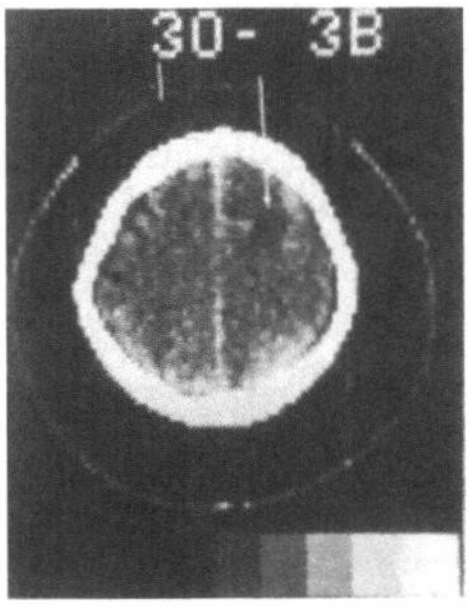

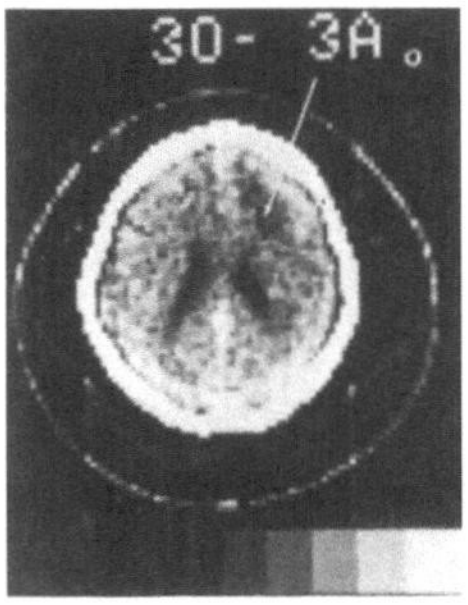

Fig. 3. This demonstrates the enhancement obtained by injection of methylglucamine iothalamate. Note how the lesion seen in the right frontal lobe is considerably sharper in outline than the same lesion which was seen in Fig. 2, Case 15

According to RUSSELL and RUBINSTEIN (5), glioblastoma multiforme tends to expand the brain, is spherical when it is in a single lobe and irregular when it covers several lobes. Cysts with clear yellow fluid as well as hemorrhage and necrosis are common in glioblastoma multiforme. Accordingly, as stated by NEW et al. (4), one would expect not only wide variation between tumors but within the same tumor. This was our experience as well. Our four cases of glioblastoma multiforme are shown in Fig. 4. Compared to the lower grade astrocytomas, they are generally of a less homogeneous lucency. Cases 302 and 222 particularly illustrate the variation within the same tumor. The surgeon's description of the tumor in Case 222 was that it was "a soft grayish well-vascularized tumor. It was removed partly with biopsy forceps and aspirated partly with a sucker. The tumor was infiltrating, had areas of necrosis and gave the impression that it was a well-vascularized malignant tumor." The gross pathologic description was of grayish-tan tissue and necrotic blood. The tissue revealed necrotic friable areas as well as firm slightly fibrotic appearing areas. No calcifications were encountered. The average NAVs within the tumor in the four cases are illustrated in Table 3 together with the average value for normal gray matter in the same scans.

There seems to be a remarkable consistency in the NAV values for the glioblastomas but, of course, not enough cases have been examined to predict the degree of certainty with which the NAV value will relate to the histologic diagnosis.

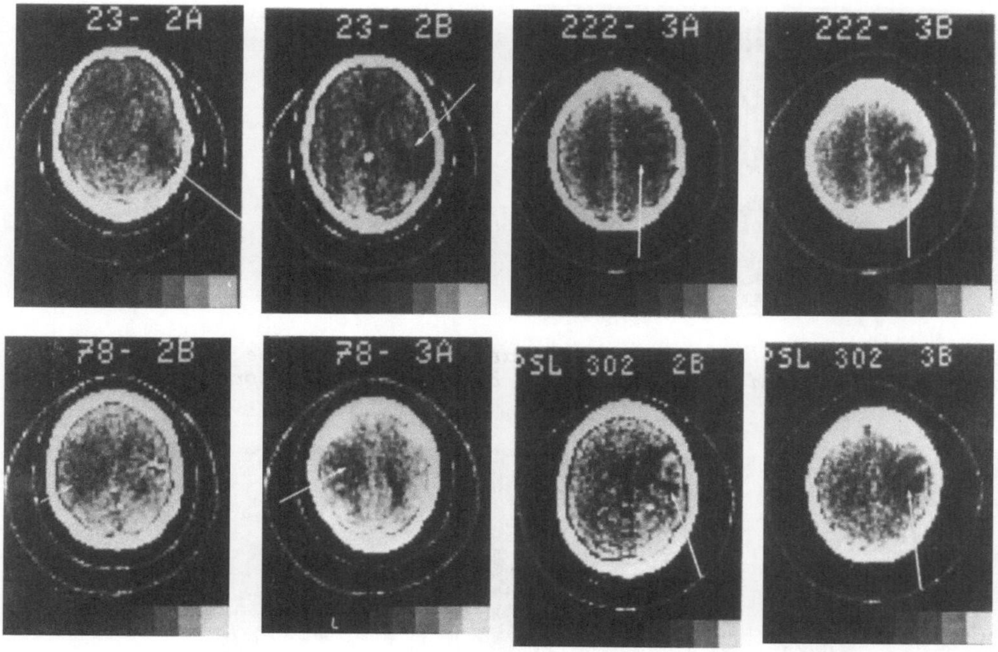

Fig. 4. Case 23 shows a right posterior temporal glioblastoma multiforme. Case 78 shows a left parietal glioblastoma multiforme. Cases 222 and 302 show right parietal glioblastomas. Note the variability within the tumors, particularly in 222 and 302

Table 3. Average NAV for tumor and normal gray matter in the same scan in cases of glioblastoma multiforme

Case	Avg NAV of tumor	Average value of normal gray matter
23	12.0	18.0
78	12.45	18.1
222	12.04	19.0
302	13.1	19.5

Our series also contained two medulloblastomas. One of these had a NAV of 21.2 and was in the posterior fossa in the midline (Fig. 5). It was of a rather homogeneous quality and had no surrounding edema. However, our experience up to that time indicated that this value probably represented a meningioma. Other neuroradiological studies were performed at another institution and did, indeed, indicate a posterior midline lesion. However, at surgery it was found to be a medulloblastoma. According to RUSSELL and RUBINSTEIN (5), medulloblastomas are usually soft, friable and well-demarcated from normal soft tissues. Although necrosis may occur in some of the very large ones, cystic changes and calcification are rare. This pathologic description does not seem to give any clue as to the nature of the CT picture in this lesion. However, it does seem safe to assume that a medulloblastoma can have a NAV

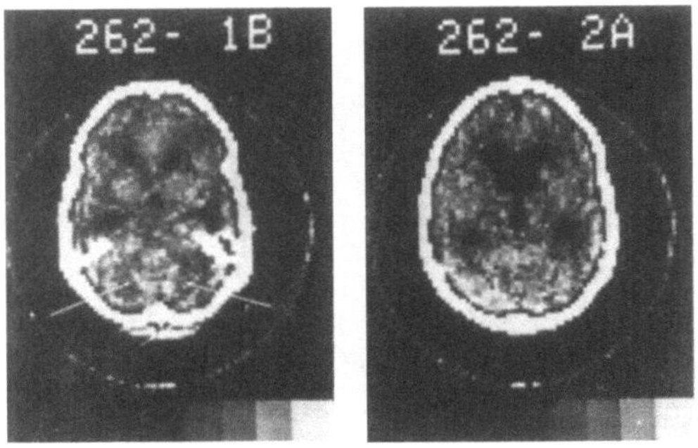

Fig. 5. This shows an area of increased density in the midline of the posterior fossa which turned out to be a medulloblastoma

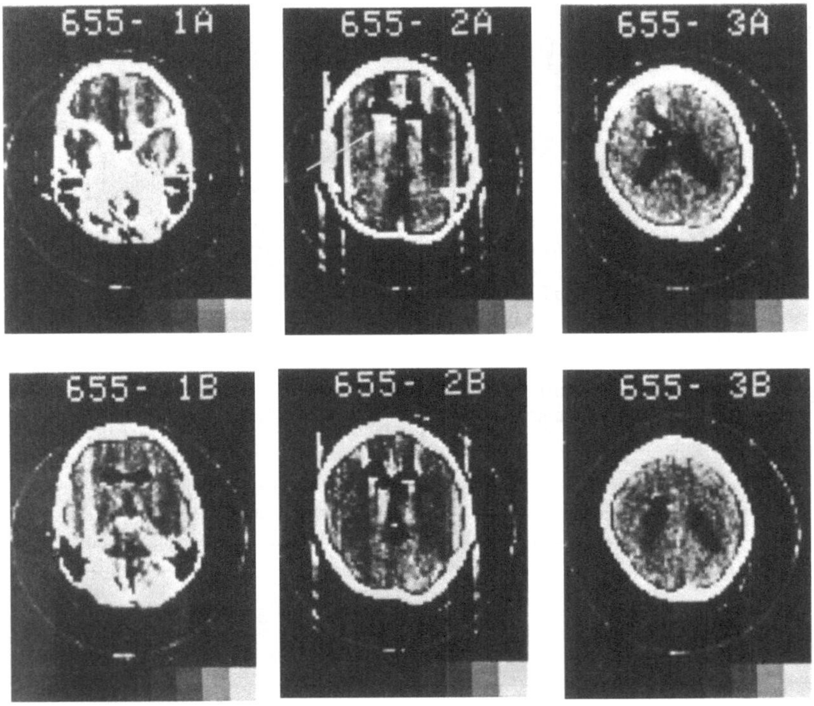

Fig. 6. This shows the artifact of computer overswing due to the presence of air in the frontal horns. As a result, it was almost impossible to clearly delineate the area of the Foramen of Monro

somewhat higher than normal soft tissue. The other medulloblastoma occurred in a young man who had had surgery and radiation therapy for a posterior fossa medulloblastoma fourteen years ago. The CT scan on this patient was performed after pneumoencephalography. As a result, there was air in the ventricular system (Fig. 6). Air in the ventricles is often responsible for a computer overswing artifact. The lesion did not visualize in the area where it would have been expected perhaps because of air in the ventricles. However, in spite of the pneumoencephalographic picture which was known at the time the CT scan was performed, it was still diagnosed as a normal CT scan. Perhaps had the air been absent, we might have been able to detect the lesion in the area of the Foramen of Monro.

CONCLUSION

Computerized axial transverse tomography (CT) appears to be a potentially useful agent in the diagnosis of gliomas as well as other intracranial masses. The CT scan will usually present the glioma as an area of decreased density when compared to the numerical absorption values of gray and white matter which are 18 and 16 units respectively. The masses may also be seen on the CT scan to displace the pineal gland and various parts of the ventricular system. In our series of fourteen known gliomas which appeared in our first 1100 CT scans, we were able to make the diagnosis of an intracranial mass in thirteen out of the fourteen cases. The one false-negative scan had an obscuring technical artifact. Nine of the thirteen cases were put in the category of glioma or metastasis which we considered to be a positive diagnosis. The oligodendrogliomas in the series were all diagnosed incorrectly in terms of their histology. However, they all contained either a great deal of calcification or hemorrhage which tended to obscure the underlying abnormality. All of the Grade II and III astrocytomas and glioblastomas multiforme were correctly diagnosed. The lower grade lesions appeared to be somewhat more homogeneous than the glioblastomas. Glioblastomas tended to show more evidence of variation within the lesions suggesting cyst formation and necrosis. One medulloblastoma was incorrectly diagnosed presumably because of technical artifact on the scan. The other medulloblastoma was misdiagnosed as a meningioma because of an apparently mistaken assumption that its numerical absorption value was too high for a glioma. In one case of a low grade astrocytoma, the diagnosis was made by repeating the scan following the administration of iodine-containing contrast material. This type of enhancement promises to further improve our accuracy in diagnosing gliomas.

REFERENCES

1. AMBROSE, J.: Computerized transverse axial scanning (tomography): Part 2. Clinical application. Brit. J. Radiol. 46, 1016 (1973).
2. BAKER, H. L., JR., CAMPBELL, J. K., HOUSER, O. W., REESE, D. F., SHEEDY, P. F., HOLMAN, C. B.: Computer assisted tomography of the head. Mayo Clin. Proc. 49, 17 (1974).
3. HOUNSFIELD, G. N.: Computerized transverse axial scanning (tomography): Part 1. Description of system. Brit. J. Radiol. 46, 1015 (1973).
4. NEW, P. F., SCOTT, W. R., SCHNUR, J. A., DAVIS, K. R., TAVERAS, J. M.: Computerized axial tomography with the EMI scanner. Radiology 110, 109 (1974).
5. RUSSELL, D. S., RUBINSTEIN, L. J.: Pathology of tumours of the nervous system, Third Edition. London: Edward Arnold (publishers), Ltd. 1971.

Heavy Ion (Proton) Radiography in Medical Diagnosis

W. V. STEWARD

INTRODUCTION

In 1895 ROENTGEN discovered x-rays, and now, after nearly eighty years, we see how sophisticated and invaluable the technique of x-radiography has become in detection including the revealing of cerebral gliomas.

However, it is worth reflecting that according to our present state of physical knowledge, we are allowed essentially three "windows" by which we may look upon the world. One of these, extensively used in diagnosis, is the electromagnetic spectrum typified by x-rays, gamma rays, and infra-red waves. Another is what may be termed a "mechanical" window typified by sound and ultra-sound in which the techniques are steadily being developed. The third and final "window" is that which uses sub-atomic particles, particularly protons and heavy-ions, and somewhat surprisingly not until now has this been used for medical diagnosis.

This report will give the first results obtained in this area using protons and will also outline those properties of protons which allow the production of unusually high contrast radiographs. Further, as the method is still in its formative period, some predictions on its potential and future development will be made.

PHYSICAL PROPERTIES OF PROTONS AND THE PRODUCTION OF HIGH-CONTRAST RADIOGRAPHS

With regards to the physical characteristics of protons the flux-depth curves illustrated in Figure 1 compare the difference in transmission of a beam of monoenergetic protons with a beam of x-rays in the diagnostic energy range when both are passed through a homogeneous absorber (1).

As can be appreciated, there is little attenuation of protons until the end of their range. Thus, a small change in effective absorber thickness will result in a relatively large change in the number of particles which penetrate, and in the energy transmitted. This contrasts with the exponential, or near exponential fall-off of x-rays and neutrons, and for a given input allows for significantly more useful informational content in the downstream signal. Figure 2, a depth-dose curve for protons in water, illustrates a large maximum, the Bragg Peak of ionization, toward the end of the particle range. This sharp cut-off and general distribution which is also important for high-contrast radiography, is again substantially different from the exponential dose distribution

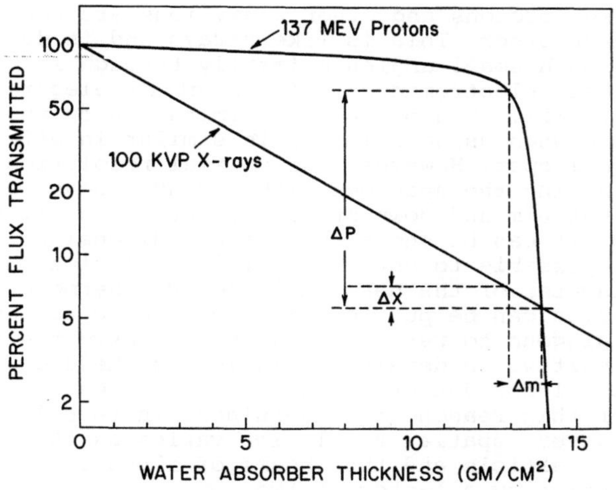

Fig. 1. Flux-depth curves for protons and x-rays passing through a homogeneous medium. Note the difference in available contrast between the two rays for the detection of mass differences

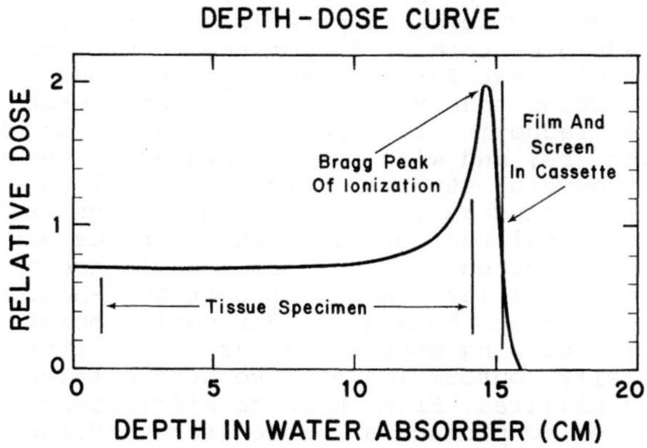

Fig. 2. Dose/depth curve for 148MeV protons in water showing the Bragg Peak of ionization near the end of the particle range. The position of the tissue specimen and of the film and intensifying screen relative to the Bragg Peak are indicated

seen with x-rays. Further, it is to be noted that the tissue specimen may be placed upstream to the Bragg Peak such that the dose throughout the depth of the specimen is near to that at the first surface.

Another important difference between x-rays and protons lies in the relative insensitivity of proton energy loss, i.e., transmission, to the chemical composition of the absorber. Thus absorber (tissue) density and hydrogen content are the main parameters which slow down protons, and whereas the mass absorption coefficient of x-rays in the diagnostic energy range varies approximately as Z^2 (Z = atomic number), for protons, this dependency is roughly proportional to $Z^{1/4}$.

Also, protons and other heavy ions are scattered during passage through an absorber. This is like x-rays and is due to repeated deflection through small angles primarily through interaction with the Coulomb field (electromagnetic field) of the electrons and nuclei of the target material. This produces a limitation on the spatial resolution obtainable when using film and is similar in effect to working with a large focal spot. However, the spatial resolution can be greatly improved by scanning the specimen with a pencil beam, e.g. 1mm in diameter, using up-stream and down-stream detectors on either side of the specimen. Also, it can be shown that with 3-dimensional reconstruction methods it is possible to obtain spatial resolution which is less than that of the diameter of the beam employed. Furthermore, this multiple Coulomb scattering can be put to good use to show up edges of internal structures analogous to xeroradiography (2). Nevertheless, this method is not so sensitive to density variation as the end-of-range technique, nor so useful for detecting density anomalies with ill-defined boundaries. For this reason, its usefulness in revealing lesions may be restricted. However, spatial resolution varies as the square root of the mass of the particle and thus beams of alpha particles should increase the resolution by a factor of two.

It is clear that the ideal diagnostic tool is the one which maximally visualizes and differentiates the bodily tissues with the minimum of hazard to the patient. This raises the question of radiation dose with protons (and other heavy ions), not only with respect to the physical dose delivered but also taking into consideration the biological effect of the particles compared with x-rays. It is apparent from what has already been said, and from the results of our experiments (see illustrations), that for a given input protons can convey considerably more information than x-rays. Furthermore, proton radiography offers a tremendous reduction in dose. It is not clear what this level of reduction might be, and will vary according to the amount of information required, but several studies indicate it will be very considerable (3, 4, 5). Thus, using an imaging system, proton radiographs have been taken of a biological test object with a surface dose of approximately 1×10^{-4} rad-equivalent to the normal environmental radiation experienced by an individual (at sea level) over an 8-hour period (6). Also, and more recently at the Argonne National Laboratory Zero Gradient Synchroton with scanning proton radiography of fresh human brain specimens, isodensity contour images have been obtained with a dose of approximately 1.7 millirad. From these considerations, it is clear that the dose required to produce an image is considerably less than those obtaining with x-radiography. This not only suggests a large safety factor when used for diagnosis in the patient per se, but a potential for screening those at risk. This is enhanced when it is recalled that the relative biological effect (RBE) of protons compared with x-rays is less then 1 (7), over a wide energy range and for a wide variety of biological test systems.

PRESENT WORK

To-date, our investigations have centered upon visualizing a variety of lesions in human autopsy and surgical pathological material and in a living patient (1, 3, 8, 9). These have included the visualizing of intracerebral hemorrhages, sub-dural hematomas, cerebral infarctions, and the lesions of multiple sclerosis. Work has also been performed on tumor-bearing human brain specimens, and as gliomas form the subject of this Symposium, these results will be described.

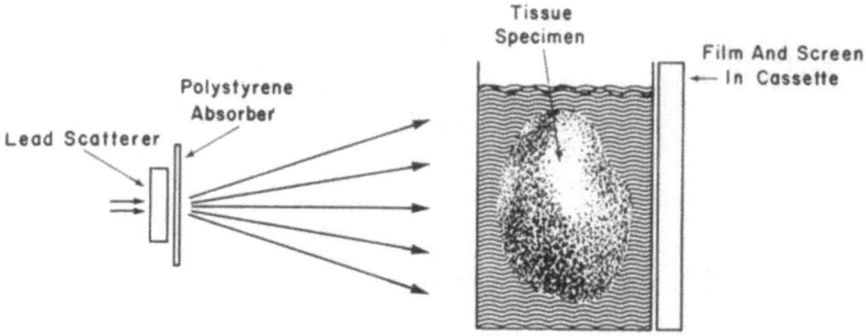

Fig. 3. Demonstration of the experimental set-up for study of specimens. Uniform illumination of the specimen is obtained by transmission of the beam through a lead or copper scatterer. As tissue density variations rather than shape variations are of primary interest, the entire specimen is immersed in water or other tissue-equivalent fluid contained in a plastic box with parallel faces. Photographic film (Polaroid TLX) is placed immediately against the downstream face of the water box in the region of the far downward slope of the Bragg Peak. Thin polystyrene absorbers are then used to adjust the energy of the beam in order to obtain radiographs of optimum constrast

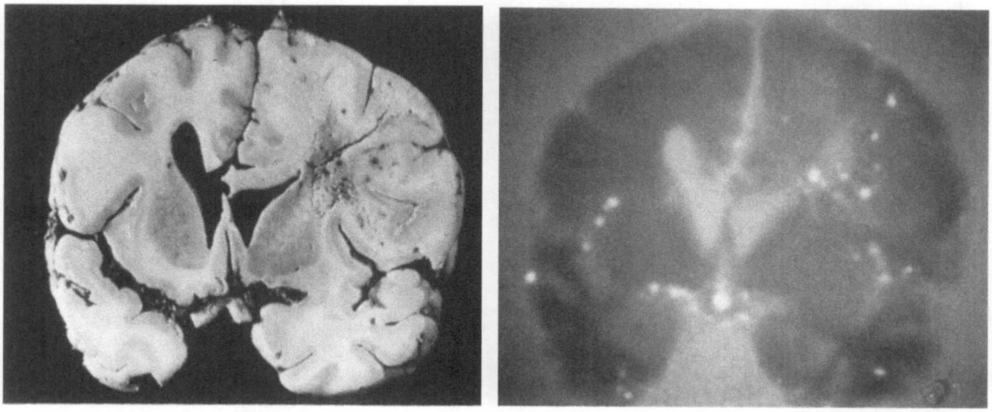

Fig. 4.(a) Front view of a coronal slice 1cm thick of a human brain fixed in formalin. Note the tumor, a glioblastoma multiforme, in the white matter of the hemisphere on the right, with swelling of the hemisphere and distortion of the ventricular system, (b) Proton radiograph (positive). Note the decreased density in the tumor area and the visualization of the basal ganglia. The white spots are air bubbles. The conditions were 115cm focus to film, no screen, Polaroid TLX film.

Initial experiments utilized the 160MeV proton beam from the Harvard Cyclotron; figure 3 demonstrates the set-up employed.

As can be appreciated from the Figures (4, 5), both primary and secondary brain tumors are well visualized by proton radiography -- a result in striking contrast to those obtained with x-radiography even under optimal conditions. Also, with proton radiography, it is possible to see the outlines of both the hemispheres and ventricular system and to visualize the basal ganglia (Fig. 4). Furthermore, as shown in Figure 6, it is possible to obtain a direct recording of a cerebral tumor within the skull despite the modulation of density due to variable thickness

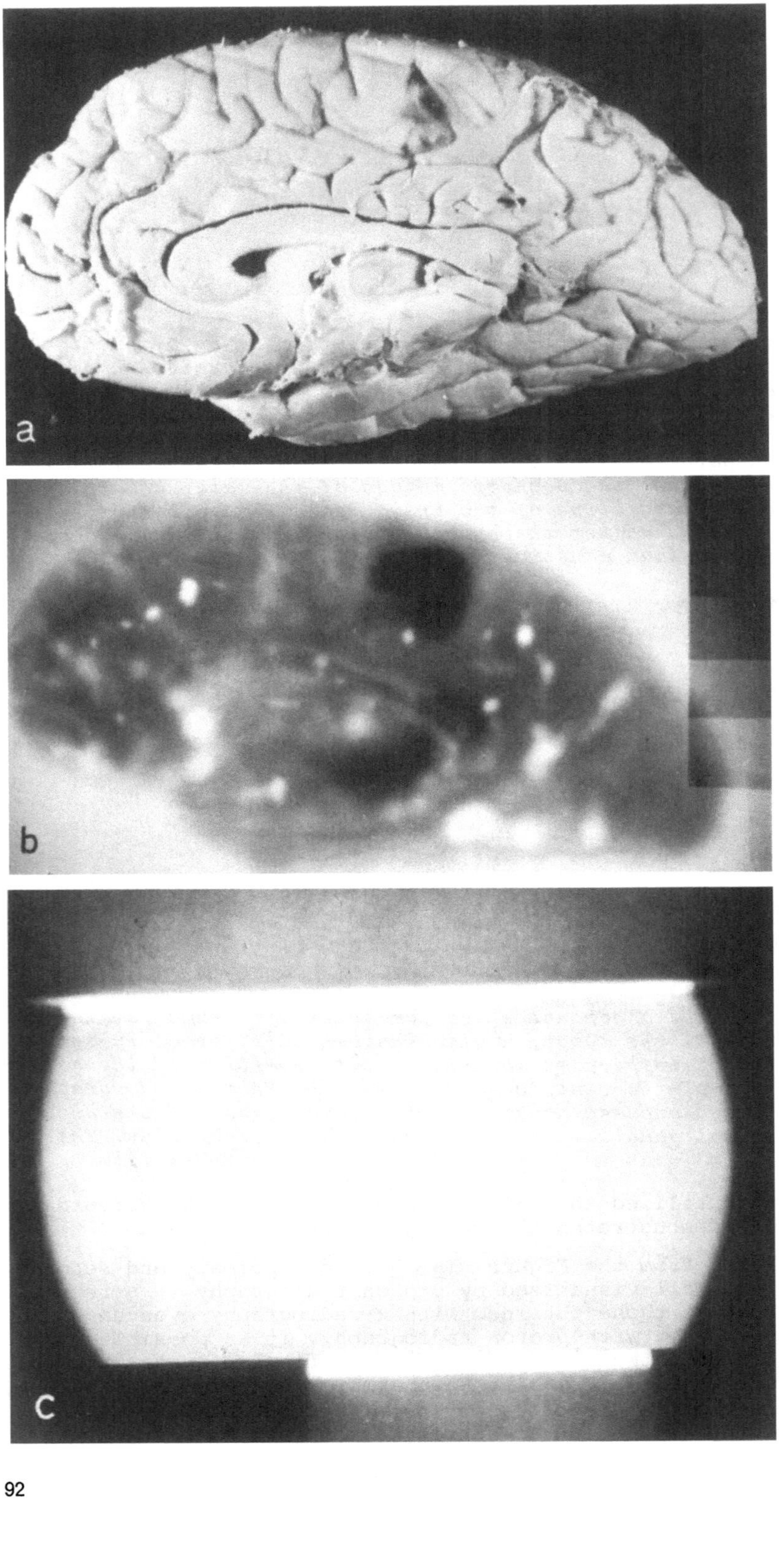
a
b
c

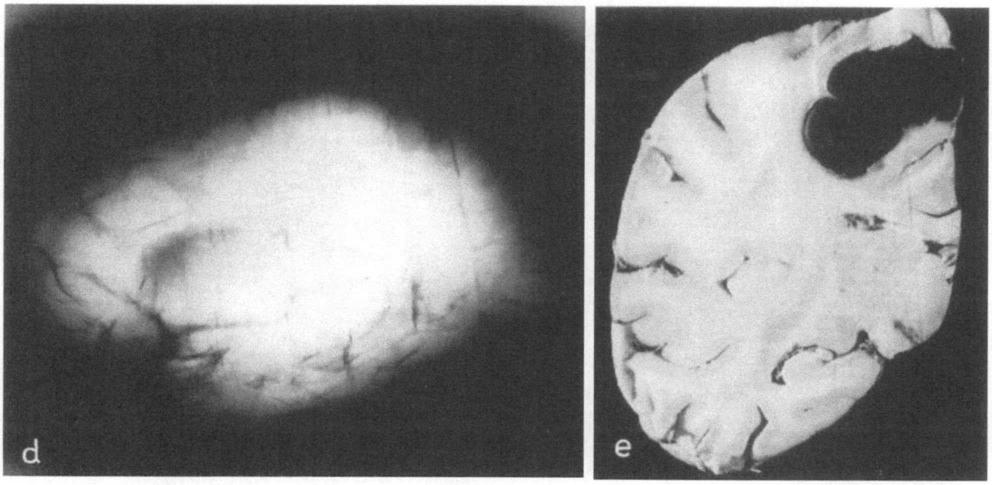

Fig. 5. (a) Medial view of formalin-fixed right hemisphere, showing a metastasis from a mucin-producing adenocarcinoma of the pancreas below the midpoint of the superior border, (b) Proton radiograph of the hemisphere (positive). The conditions were 115cm focus to film, no screen, Polaroid TLX film, (c) X-radiograph of the hemisphere in a plastic water box. The conditions were 126cm focus to film, Du Pont high-speed 1A screen, Polaroid TLX film in Polaroid cassette, 30 KV peak, 200MA, 3.0 seconds, (d) X-radiograph of the hemisphere in air. The specimen was reconstructed from serial coronal sections. Note that the tumor is just visible. The conditions (optimal) were 92cm focus to Kodak mammography film, 27KV constant potential, 20MA, 2 minutes, (e) Photograph of a slice taken through the tumor

in the cranial bones (9). The situation is essentially equivalent to the intact human head and, as far as we are aware, this is the first time that a noncalcified brain tumor has been directly recorded on film.

However, film as the recording medium for radiography has a number of inherent limitations, for instance, limited information capacity and dynamic range. Electronic imaging systems would not only overcome these limitations but importantly would also a) improve the spatial resolution, b) allow the trading off of spatial resolution versus residual proton energy resolution ("levels of gray"), c) potentially have quantum efficiency and thus minimize the dose to the patient, d) make feasible the viewing of organs without their immersion in fluids, and e) by drawing off digitized information, permit image processing, including three-dimensional reconstruction. For these reasons, therefore, it is important to explore these alternatives to the film method.

Nevertheless, scanning proton radiography has many advantages, and, supported by National Cancer Institute contract NO1 CB-43918, is actively being pursued in a collaborative effort between the University of Chicago and members of the Accelerator Research Division, Argonne National Laboratory.

As a preliminary first step, and in order to obtain a better understanding of the physics and engineering problems involved, a secondary beam of 200 MeV monoenergetic particles (designated beam 42) was obtained

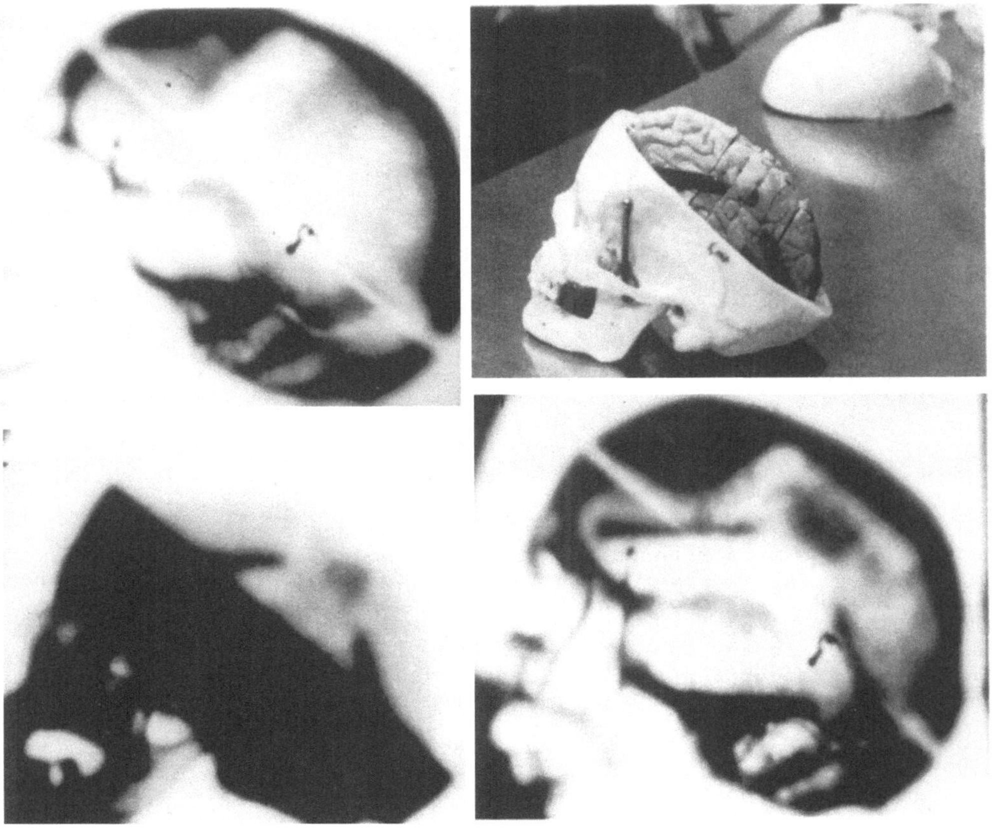

Fig. 6. (a) Proton radiograph (positive) of normal formalin-fixed human brain autopsy specimen placed within a skull in a water tank. Comparison with a similar picture taken with the brain removed (not shown) indicates some areas of increased density coming in from the periphery. Polaroid TLX film. Radelin TF intensifying screen. First surface dose less than 0.5 rad, (b) Photograph of the right hemisphere of a human brain (specimen in Figure 5) placed within the lower portion of a skull. Two strips of lead attached to the brain point towards a small portion of the tumor extending to the medial surface of the hemisphere, (c) Proton radiograph (positive) of the tumor-bearing hemisphere held in the lower portion of the skull as illustrated in Fig. b. Note the ends of the lead markers pointing to the lesion. Polaroid TLX film. Radelin TF intensifying screen. First surface dose less than 0.5 rad, (d) Proton radiograph (positive) of the brain and tumor in the complete skull. With the skull cap in place, despite the modulation of density due to variable bone thickness, the tumor density can still be seen. Polaroid film. Radelin TF intensifying screen. First surface dose less than 0.5 rad

from the 12GeV Zero Gradient Synchrotron (ZGS) sited at the Argonne National Laboratories (10). With this set-up, a fresh autopsy brain specimen with an intra-cerebral hemorrhage was compared with a normal fixed brain specimen. Despite the relatively low resolution used, much information was obtained and, even allowing for differences that are to be expected in density and contouring between fresh and fixed specimens, it was clear that the intracerebral clot was strikingly revealed.

Further work is now in progress to push the scanning method to a higher level of sophistication and for this the output of a synchrotron, used as a booster to inject 200 MeV protons into the Argonne Zero Gradient Synchrotron, will be employed (11). Attached to the synchrotron is a collimation system under construction for scanning with arrays of 1mm diameter beams. This will be ready within one month of this writing, is approximately 60 feet in length, and will be employed to rapidly scan specimens on an automatic stage.

PROSPECTS FOR THE FUTURE

Historical precedent indicates that the development and prospects for heavy ion radiography are best compared and are closely linked with those of x-radiography. But whereas the latter has had many decades to evolve, heavy ion radiography has been in existence for a mere two and one-half years. Thus, though advances in this new form of radiography have been considerable, we still have some way to go.

It is, therefore, but early days for heavy ion radiography, and despite the promise of the experiments done so far, it has yet to prove itself a useful diagnostic tool. For this to be accomplished, it is crucial that the technique be brought to the patient in the hospital and that it is feasible to design an accelerator suitable for in-hospital use. Fortunately, this is eminently possible. According to Dr. R. MARTIN, Director of the Accelerator Research Division, Argonne National Laboratory, such an accelerator would be a synchrotron consisting of a ring of magnets arranged in 21-foot diameter circle. This would cost $ 100,000 - $ 200,000, and be simple and reliable compared to machines used for physics research. Importantly, it could also look at six patients simultaneously.

It may be argued that ions heavier than protons would be more useful than protons themselves for radiography, their lesser scattering giving greater spatial resolution and hence improvement in the image quality. But this must be weighed in the context of dose, and practical reality. To obtain perspective in this area, calculations have been performed for a number of ions (from protons to argon) to determine the minimum dose and size of accelerator required to detect a given density or thickness change in a biologically appropriate situation (12).

Except for deuterium and tritium ions, even under ideal conditions, the dose for the heavier ions is greater than for protons and their cost of production very noticeably increased. Furthermore, computational methods may improve the spatial resolution well below the diameter of the scanning beam and these plus the price to be paid for the additional informational content is currently being explored by the writer. Altogether, therefore, the case for using the heavier ions is presently rather questionable but, as with most things, will only be decided in the light of further experience, knowledge, and by clinical need.

It must be kept in mind that the objective throughout is to produce images of the highest diagnostic quality and for this the wealth of information accumulated by the physicist has given us a very firm theoretical foundation. In fact, even now we have an equal understanding of how heavy ions pass through an absorber as we do with x-rays. Thus, most of the factors which determine the absorption of protons are well known and are capable of being measured. Importantly, it must also be remembered that heavy ion radiography conveys information of a radically different kind from that of x-rays, and that in the production of images many intangibles are involved (13, 14).

Nevertheless, in the specific case of the cerebral gliomas, it is anticipated that these will, unlike with conventional x-radiographic procedures, be directly as opposed to indirectly visualized. Hopefully, also, the sensitivity of the method should allow the differentiation of the various tumor types, the visualization of secondary involvement of adjacent structures, and confirm the presence and perhaps the degree of cerebral oedema. Moreover, this may be accomplished with a wider safety margin than has heretofore been possible and with minimum disturbance to the patient.

Overall, therefore, it is felt that heavy ion radiography will contribute significantly to diagnostic medicine and that its usefulness will be comparable to that presently seen with x-radiography.

Acknowledgements

Supported by National Science Foundation Grant GI 32991X, National Cancer Institute Contract #NO1-CB43918, Argonne National Laboratory, Argonne, Illinois, work project #3144, and by a donation from the Harry Rush Memorial Fund.

REFERENCES

1. STEWARD, V. W., KOEHLER, A. M.: Proton radiographic detection of strokes. Nature 245, 38 (1973).
2. WEST, D., SHERWOOD, A. C.: Radiography with 160 MeV protons. Nature 239, 157 (1972).
3. STEWARD, V. W., KOEHLER, A. M.: Proton radiography in diagnosis of breast carcinoma. Radiology 110, 217 (1974).
4. STEWARD, V. W.: Proton radiography in the detection of tumors and strokes. In: Argonne National Laboratory Symposium on Advanced Applications of High Energy Physics, p. 3.1, May 17 (1973).
5. Exploration of the use of a proton beam in tissue densitometry. National Cancer Institute proposal contract no. NO1 CB-43918.
6. GOTTCHALK, E., BOYD, D., KOEHLER, A. M.: Private Communication.
7. UENO, Y., GRIGORIEV, Y. G.: The RBE of protons with energy greater than 126 MeV. Brit. J. Radiol. 42, 475 (1969).
8. STEWARD, V. W., KOEHLER, A. M.: Proton beam radiography in tumor detection. Science 179, 913 (1973).
9. Steward, V. W., KOEHLER, A. M.: Proton radiography of a human brain tumor within the skull. Surg. Neurol. 2, 283 (1974).
10. COLTON, E.: Accelerator Research Division, Argonne National Laboratory. Private Communication.
11. MOFFETT, R.: Accelerator Research Division, Argonne National Laboratory. Private Communication.
12. KOEHLER, A. M., GOITEIN, M., STEWARD, V. W.: Calculation of minimum dose required to detect a thickness or density anomaly in charged particle radiography. To be published.
13. ROSSMANN, K., WILEY, B. E.: The central problem in the study of radiographic image quality. Radiology 96, 113 (1970).
14. ROSSMANN, K.: Point spread function, line spread function and modulation transfer function; tools for the study of imaging systems. Radiology 93, 257 (1969).

Treatment of Supratentorial Gliomas

P. C. Bucy

It would be a pleasure to appear before you, able to outline an effective surgical treatment, or for that matter, any effective treatment for gliomas of the cerebral hemispheres. Unfortunately that is not the case. And it would not be a grave error at this point for me to admit my incompetence and retire.

Just 90 years ago the first glioma of the cerebrum was diagnosed on the basis of its neurological manifestations by ALEXANDER HUGHES BENNET and operated upon and removed by RICKMAN GODLEE with the assistance of the noted cerebral physiologist, DAVID FERRIER. This was a major accomplishment for 1884. The patient survived the operation, but unfortunately he did not survive his tumor. Shortly after the operation the tumor began to grow again. It soon produced more evidence of neurological disorder. And the patient died.

For the remainder of the 19th century not only operations upon gliomas of the brain but all neurosurgical procedures fared badly. The surgical mortality of all operations upon the brain fluctuated between 50 and 100% depending largely upon who undertook to operate. Those who did so included MACEWEN in Scotland, HORSLEY in England, KRAUSE und VON BERGMAN in Germany and KEEN in the United States. Because of this appalling mortality, neurological surgery was rapidly losing the favor that the initial operation brought and it might well have expired except for an energetic, determined, extremely competent and superbly trained young man in the United States - HARVEY CUSHING. CUSHING correctly recognized and accepted his responsibility. He was determined to demonstrate that surgery upon the brain could be performed with an acceptable operative mortality. And this he did. When he retired in 1932, about a third of a century later, his surgical mortality for all brain tumors was 6.8% and for gliomas was 11%. This was a remarkable accomplishment, but by no means all that CUSHING had done. He had taught the medical profession how to recognize the different types of brain tumor, he had shown surgeons how to operate upon them without killing the patients, and he had trained most of the neurosurgeons of the world.

But there was one thing that CUSHING did not do, and this he was the first to recognize. He did not learn how to cure gliomas of the cerebrum. And neither has any one else in the many years since then. But advances have been made. We have further lowered the operative mortality, have prolonged the patients' lives and improved their conditions following operation. It appears that we have been able to cure an occasional patient. The advances which have been made rest upon a number of decisive factors. First, the general medical profession has been educated as to the common early manifestations of brain tumors and as a result, patients suspected of harboring such tumors are referred much earlier to the neurosurgeon than they formerly were. As a result, many patients are now seen by the neurosurgeon when there are few or no neurological abnormalities and when the tumor can neither be diagnosed nor localized on the basis of a history and a neurological examination. Fortunately, our diagnostic armamentarium has kept pace. Pneumoencephalography and electroencephalography were developed and have now been replaced by more accurate technics. Cerebral angiography and radioactive brain scans have made it possible to recognize the presence of a tumor much earlier than was formerly the case, to localize it accurately and in some instances to determine its pathological nature. A little over a year ago the EMI scanner was announced. This extraordinarily sophisticated computerized X-ray technic makes it possible to diagnose and localize tumors and other cerebral lesions without the discomfort of cerebral angiography, without injecting any radioactive material or anything else. In addition, this technic is also capable of providing information regarding the density of the pathological lesion. This, eventually, should make it possible to make a pathological diagnosis before the patient is operated upon, in many instances.

CUSHING recognized many years ago that gliomas of the brain varied widely in their pathological nature. He also recognized that one could not deal intelligently with all gliomas as one group, but must treat the various kinds of gliomas differently. Accordingly, he provided PERCIVAL BAILEY with the opportunity to study his collection of gliomas, to classify them and to demonstrate the clinical correlations of the various types. Nevertheless, this information has not proved as beneficial as had been hoped. In fact it, for a time, seemed to lead us into an era of inactivity so far as the treatment of the more malignant gliomas, the glioblastomas, was concerned.

In general, it can be said that the only satisfactory treatment of a cerebral glioma is to remove it. Whenever it is possible to remove a glioma completely that is true, but all too frequently safe, complete removal is not possible. Gliomas which occupy the occipital or the frontal or temporal poles of the cerebral hemisphere can at times be completely removed and the patient cured (HALSTEAD, WALKER and BUCY), but such cases are all too few. Cystic astrocytomas of the cerebrum can on occasion be cured (ELVIDGE). Cystic ependymomas are even more rare, and they too can be cured.

Most patients with astrocytomas and oligodendrogliomas of the cerebrum give evidence in one form or another that the tumor has been there for many years. The arrival of the patient on the neurosurgical service is commonly the result of sudden increase in symptoms. It is my belief that in many cases this recent acceleration of symptoms is the result of malignant changes occurring within the tumor and a resulting acceleration of its growth. As a result the improvement which occurs following the removal of the tumor does not persist as long as the protracted history of the case would lead one to expect. This is not to deny that on occasion the results of removal of an astrocytoma or an oligodendroglioma are most gratifying.

In general, the neurosurgeon should adopt the rule that all readily accessible gliomas should be removed as completely as possible. Even some which are not as accessible should be similarly dealt with. The famous patient that HARVEY CUSHING operated upon on May 6, 1932 before the assembled founders of the Harvey Cushing Society married shortly after the operation, had two children and was well and living when last heard from over 13 years later (FULTON). She had a glioma (an astroblastoma) in the wall of the third ventricle. Although this is not an isolated case, in general it is not likely that gliomas involving the thalamus or the basal ganglia will often be removed with such favorable results.

For many years it was generally felt that malignant gliomas could not be successfully dealt with and should be left alone once the diagnosis had been made. As a result some surgeons having exposed such a tumor would take a small piece for microscopic examination and if their suspicions were confirmed they would remove a part of the skull or make a decompression and leave the tumor alone. Other surgeons would insert a needle into the tumor through a drill hole and do nothing further if the tumor proved to be a glioblastoma. Still others would rely upon the appearance of the angiogram for the diagnosis and would not operate upon glioblastomas at all.

In my opinion all of these defeatist attitudes were wrong. I recognize that it is seldom that one can cure a glioblastoma - although I think we have done so in a few cases. The important point is that to operate upon a glioblastoma and not remove the tumor results in a high surgical mortality - 50% - and those patients who survive the operation steadily deteriorate. If, on the other hand, the tumor is radically removed and dexamethasone is given, the operative mortality can be reduced to less than 3%. Even more important, the condition of the patients will improve as compared with their preoperative state and as compared with those patients in which this policy is not followed. Furthermore, their condition following the operation will be much better than if only a biopsy is made or if only a limited extirpation of the tumor is undertaken. With radical extirpation, 13% of the patients will be alive two years and 3.5% five years after operation, whereas with less extensive removals all patients will be dead within 12 months. There is another group of patients with glioblastomas who do even better. These patients have tumors which appear more circumscribed, are often adherent to the dura mater and frequently are found in the temporal region. These tumors form about 20% of all glioblastomas. With radical removal, 11% of these will be alive five years after operation and a few have gone 8 to 15 years in good health. Many a family have expressed their appreciation for the added years of health and life which such radical extirpation of these tumors has afforded.

Frequently one hears the remark that such radical extirpation of the tumor should be considered only when the tumor is in an area of the brain where its removal will not cause further neurological disability. This policy will deny to many patients the benefit that radical extirpation can bring. Even when the tumor is located in the so-called motor area, or where it is producing a partial or complete aphasia it must be borne in mind that the tumor itself is not capable of neural function and its removal will not produce further neurological deficit. One must also remember that in most instances the tumor will push functioning areas of the brain aside while impairing function by compression. As a result in 60% of the cases the patient's neurological condition will be better following a radical extirpation than it was before operation.

There has been considerable controversy as to the value of radiation therapy of glioblastomas. The reason for this is not difficult to learn. The benefits of radiation therapy are small. Nevertheless, they are there, even though quite limited. The benefits of radiation therapy do not extend beyond two years. However, in a group of patients selected at random those who received radiation therapy lived longer and the quality of their survival was better during that two year period than those not so treated. Accordingly, it is my opinion that in addition to radical surgical removal, all glioblastomas should be given radiation therapy shortly after they have been operated upon. They should receive 4,500r in divided doses of 200 to 250 r given daily, five days a week for about a month. Unfortunately no one has exhibited sufficient interest in the matter to determine whether this maximum dose administered in this fashion gives the optimum results.

Unfortunately a second course of radiation therapy is seldom sufficiently beneficial to justify its being given. The same is usually true of reoperation upon these malignant gliomas.

The more benign gliomas of the cerebrum - the astrocytomas and oligodendrogliomas - should also be treated by radical extirpation. However, radiation therapy should not be given unless microscopic examination of the tumor discloses evidence of malignant changes. Admittedly this point has not been as carefully investigated as has the radiation therapy of glioblastomas. My impression is that beneficial results do not follow radiation treatment of these tumors, nor in light of the limited period during which such treatment is beneficial with glioblastomas - two years- is improvement to be expected with slowly growing tumors of this sort. Furthermore, it is becoming increasingly apparent that radiation therapy delivered to the central nervous system results, over long periods of time, in deleterious effects too serious to warrant using this form of treatment with slowly growing tumors where beneficial results are questionable and long term survival is to be expected. Radiation therapy directed to the brain can cause serious atrophy (MALAMUD, BOLDREY, WELCH & FADELL), just as it can cause paraplegia when directed toward the spinal cord (BHAVILAI). It can also generate the development of other tumors, sarcomas (MEREDITH, MANDEVILLE & KAY), meningiomas (ALEXANDER), etc.

And now we come out "by the same door where in I went." There is no known effective treatment of supratentorial gliomas. In 90 years we have made no progress toward developing a cure for these tumors. Surely this is the neurosurgeons' responsibility. Surely, also, a century of experience has indicated that it is most unlikely that these tumors will ever consistently be cured by surgical means. We must look elsewhere. We must study these tumors more carefully. We must learn their epidemiology. Why are certain gliomas predominantly those of patients over 45 years of age while others affect younger people? Why can some glioblastomas be removed and the patient make a dramatic recovery and remain well for several years, only to again suddenly develop symptoms of his tumor, rapidly grow worse and die? What stimulates the development of gliomas of the brain in the first place?

It seems likely that the answers to these and other questions will be disclosed only by the technics of the basic sciences. It is unlikely that the older neurosurgeons will learn the technics of the basic sciences. If we are to fulfill our responsibilities to mankind and gain mastery over the gliomas of the brain, then we must prepare our younger neurosurgeons to do so. They must be trained in detail in one or another of the basic sciences. They must be given the time and the facilities to acquire knowledge and experience with the scientific method. It is obvious that research is a young man's game. Either he will begin to engage in serious productive research before he is 30-35 years old or he probably will never do so. He must, therefore, become involved in research during his residency training period. He must be encouraged in this. We must never begrudge him the time that this, of necessity, takes away from his clinical duties. We must stimulate him to engage in research if he is one of those few who exhibit a capacity for research. We must provide him with the opportunity and the facilities to acquire the knowledge and technics of research and to prosecute those studies which excite his interests and for which he has both the background and the capacity. This is our responsibility.

REFERENCES

1. BAILEY, P., CUSHING, H.: A Classification of the Tumors of the Glioma Group on a Histogentic Basis with a Correlated Study of Prognosis, p. 175. Philadelphia: J. B. Lippincott Co. 1926.
2. BENNETT, A. H., GODLEE, R. J.: Case of cerebral tumor. Proc. Roy. Med. and Chir. Soc. London, 1885, n.s. 1, 438; and Lancet 1884 II, 1090. (cited after G. Jefferson).
3. BHAVILAI, D.:Inadvertent destruction of the spinal cord by radiation therapy. Surg. Neurol. 2, 333 (1974).
4. CUSHING, H.: Intracranial Tumours: Notes upon a series of two thousand verified cases with surgical mortality percentages pertaining thereto, p. 150. Springfield, Ill.: Chas. C Thomas 1932.
5. ELVIDGE, A. R.: Long-term survival in the astrocytoma series. J. Neurosurg. 28, 399 (1968).
6. FULTON, J. F.: Harvey Cushing: A biography. p. 756. Springfield, Ill.: Chas. C Thomas 1946.
7. HALSTEAD, W. C., WALKER, A. E., BUCY, P. C.: Sparing and nonsparing of "amulcar" vision associated with occipital lobectomy in man. Arch Ophthalmol. 24, 948 (1940).
8. JEFFERSON, Sir. GEOFFREY: Selected Papers, p. 563. London: Pitman Medical Publishing Co. 1960.
9. JELSMA, R., BUCY, P. C.: The treatment of glioblastoma multiforme of the brain. J. Neurosurg. 27, 388-400 (1967).
10. JELSMA, R., BUCY, P. C.: Glioblastoma multiforme; Its treatment and some factors effecting survival. Arch. Neurol. 20, 161 (1969).
11. MACEWEN, WM.: Tumour of the dura mater removed during life in a person affected with epilepsy. Glasgow Med. J. 12, 210 (1879). (cited after G. Jefferson).
12. MACEWEN, WM.: Intra-cranial lesions - Illustrating some points in connexion with the localization of cerebral affections and the advantages of aseptic trephining. Lancet 1881 II, 544, 581. (cited after G. Jefferson).
13. MACEWEN, W. M.: On the surgery of the brain and spinal cord. Brit. Med. J. 2, 302 (1888). (cited after G. Jefferson).
14. MALAMUD, N., BOLDREY, E. B., WELCH, W. K., FADELL, E. J.:Necrosis of brain and spinal cord following X-ray therapy. J. Neurosurg. 11, 353-362 (1954).
15. MEREDITH, J. M., MANDEVILLE, F. B., KAY, S.: Osteogenic sarcoma of the skull following roentgen-ray therapy for benign pituitary tumor. J. Neurosurg. 17, 792-799 (1960).
16. NORWOOD, C. W., KELLY, D. J., Jr., DAVIS, C. H., Jr., ALEXANDER, E., Jr.: Irradiation induced mesodermal tumors of the central nervous system. Surg. Neurol. 2, 161-164 (1974).
17. Special Note - A new diagnostic technique - The EMI-scanner. Surg. Neurol. 1, 97 (1973).

Treatment of Infratentorial Gliomas in Childhood

E. B. Hendrick, H. J. Hoffman, and R. P. Humphreys

Tumours of the central nervous system, especially those occurring intracranially are among the most common present in infancy and childhood. At the Hospital for Sick Children in Toronto, brain tumours are present in about 2.5% of all admissions to the neurosurgical division or approximately 50 cases a year.

Brain tumours, particularly in infancy, are diagnostic problems. The developing brain has a great capacity for adaptation. Mass lesions may reach great size before producing local or generalized symptoms and signs suggesting the correct diagnosis.

Increased intracranial pressure and abnormal increase in head growth in infants, may be the only manifestation of such tumours. The general symptoms of malaise, irritability, clumsiness or even infrequent vomiting may be so varied and uncharacteristic that they may be mistaken for some relatively innocuous illness in childhood. Hence, physicians responsible for child health should be ever alert at the possibility of an underlying intracranial space occupying lesion.

Detailed history taking and physical examination should include questions concerning behaviour and intellectual change. A decrease in the school performance or sudden slowing in the attainment of developmental milestones may be the first symptoms of trouble.

Lack of symptoms particularly in the cerebral hemispheres is due both to the compensating elasticity of the child's skull with its opened sutures and fontanelle, and to the ability of the immature brain to compensate for gradual functional impairment. However, the immature brain tissue is also extremely vulnerable so that severe permanent damage may be inflicted within a relatively short time. The abnormality detected by neurological examination may indicate only the tip of the iceberg. The degree of brain damage may be much more severe.

In children most brain tumours occur between the ages of four and ten years. The small number diagnosed in infants under one year of age probably reflects the difficulty in detecting behaviour change and the difficulties of precise and finite neurological examination. Most infants with brain tumours present first with the diagnosis of hydrocephalus.

A sharp decline in the frequency of brain tumours after ten years of age is difficult to explain unless one postulates that the tumours of embryonic cell type such as medulloblastomas have declared themselves at an earlier age.

In my experience at the Hospital for Sick Children, infratentorial tumours are composed of approximately 32% astrocytomas, 20% medulloblastomas, 20% brain stem gliomas and 14% ependymomas with unusual tumours making up the remainder.

The presenting symptoms on admission were predominately headache either occipital or frontal in 67%, and vomiting in 7.5%. Incoordination, diplopia, anorexia, and decreased vision occur with a slightly decreased percentage.

The physical findings on admission show a high percentage of patients with papilledema. This occurred in 67.5% of patients. Nystagmus and wide based gait being the next most commonly found signs, such findings as hyporeflexia, facial weakness, and head tilt occurred in 10% or less.

Of the diagnostic procedures following admission to hospital, plain x-rays of the skull are probably the most important. Split sutures were found in over 80% of the children admitted who subsequently had a diagnosis of infratentorial glioma. Ventriculography rather than pneumoencephalography has been a safer procedure in my experience. In the children, of course, most diagnostic procedures had to be carried out under general anesthesia and so air ventriculography is commonly combined with positive ventriculography and arteriography. More recently nuclear scanning has been invaluable.

The characteristic obstruction of the aqueduct of Sylvius and diminution and distortion of the 4th ventricle with the tonsillar herniation through the foramen magnum are the common findings.

For the past ten years it has been the policy on the neurosurgical unit at the Hospital for Sick Children following definitive ventriculography or arteriography for infratentorial glioma, to carry out a ventriculoperitoneal shunt procedure using any of the various methods according to the requirements at the time. In our hands this has reduced the difficulties of the major operative procedure, and the post-operative mortality and morbidity. In a severely cachectic child with the long history of vomiting, the shunt procedure followed by a delay of ten days to two weeks before definitive tumour surgery reduces the risks of surgery by allowing the patient to achieve an improved pre-operative nutritional state.

The operations for infratentorial gliomas are carried out under general anesthesia with controlled respiration and with the child placed in a prone position on a specially built support to allow free movement of the abdomen and lower thorax. We favour a simple mid-line muscle splitting incision with removal of bone from the foramen magnum up the level of the transverse sinus and removal of the arch of the first cervical vertebrae. In the remainder of the techniques in tumour excision, I do not vary greatly from any other centre.

Astrocytomas of the posterior fossa in children constitute only 40% of the total astrocytoma group in our experience. They do, however, carry the best prognosis in terms of functional longevity. I might say at this time that it has been our policy on the unit to remove all the tumour possible, consistent with maintaining a functional individual. I feel this simple fact is often lost sight of in the hue and cry of the chase. Total tumour excision carries no glory if the patient is left neurologically destitute.

Ependymomas are predominantly found in the posterior fossa and 67% of the group present in this survey were so located. Their difficulty in removal is, of course, compounded by their site of detachment and total removal in my experience has been more uncommon than common. Initially,

I felt that ependymomas should not be irradiated until there was some sign of tumour recurrence. However, my policy has changed over the years, and I now carry out routine Cobalt irradiation on all patients with ependymomas in the posterior fossa.

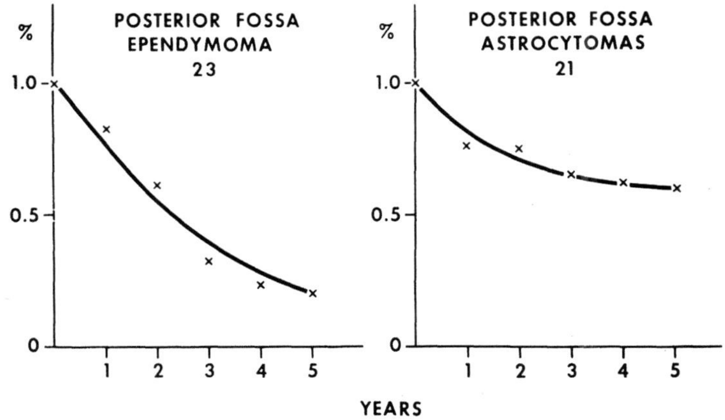

Fig. 1

The mortality statistics of the tumours showed that in spite of irradiation, the crude survival time of the ependymomas is still most unsatisfactory. Less than 25% of the posterior fossa ependymomas survived beyond a five year period. The posterior fossa astrocytomas on the other hand are not routinely irradiated, except when there is involvement of the brain stem even though the major tumour mass presents mainly in the cerebellar hemisphere. But with irradiation, the crude survival time shows that slightly better than 50% are functioning at the end of five years.

Medulloblastoma in my experience, appears to occur more commonly than as reported elsewhere in the literature. Ninety-one per cent occur in the posterior fossa, the majority occurring between the ages of four and six years. Again, in this type of tumour, I have never felt that I was able to obtain a satisfactory total removal. All these children have irradiation post-operatively.

There has been some concern about the spread of medulloblastoma during the period following the ventriculo-peritoneal shunt and indeed, following posterior fossa surgery when the tumour particles have been disseminated in the spinal fluid. We have six cases in which there has been a spread to the abdomen and from there throughout the body, as a direct result of shunting the tumour material from the lateral ventricle. At the present time, we have evolved a small implantable millipore filter in the form of a disc which can be attached in line with the main shunt mechanism and is placed in the suboccipital or lower cervical area along the line of the shunt insertion. This mechanism is under investigation but hopefully it will strain out the bulk of the gross tumour particles. It is so placed, that it receives radiation during the therapy to the posterior fossa. The spread of tumour has not been a consistent feature of the shunting in medulloblastomas and has certainly not deterred us from carrying out this valuable pre-operative maneuver.

Brain stem gliomas in children are a very distressing and, indeed, depressing problem with which to be confronted. It is apparently a general feeling amongst pediatricians and pediatric neurosurgeons that once the diagnosis has been made irradiation is the only course of treatment. I would like to differ strongly with this concept. It is obvious that the survival time of the brain stem glioma following irradiation represents a disastrous and apalling situation. While the crude survival time in the most malignant of the extra axial type of tumour, the medulloblastoma, is slightly more than 40% at the end of five years, the brain stem glioma is less than 15% at the end of five years. I would like to make a plea for a more aggressive diagnostic procedure and treatment in these unfortunate individuals.

The characteristic findings in the brain stem glioma are those of dissociated cranial nerve signs, including diplopia, and facial palsy with long track signs not necessarily related to the same body side. The history is usually short with a rapid development of severe neurological deficits. X-rays of the skull do not reveal splitting of sutures in a majority of the cases. Air study reveals thickening of the pons and the medulla with posterior protrusion of the floor of the 4th ventricle. I cannot, but agree, that this type of patient, is a candidate for irradiation only.

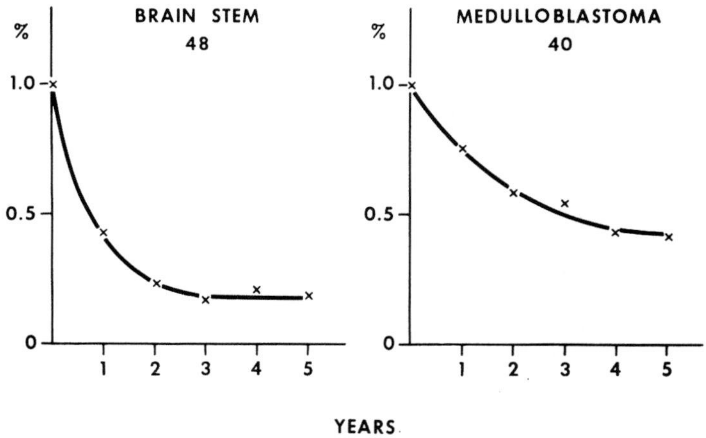

Fig. 2

There is another group of children with brain stem gliomas that present in a much different fashion and have a much different outlook. The history is usually long and there may be periods of remission and exacerbation of their neurological signs. Investigation by pneumoencephalography shows a huge mass lesion in the posterior fossa with growth expansion of the pons and medulla.

I feel these patients should have operative exploration. In my experience in the last sixteen years, I have had five patients who had large cystic lesions in the pons which had been drained satisfactorily by making a small window in the mid-line over the thinned out area tissue. One patient had soft tissue material within the depths of the pons as well as cyst fluid and this tumour material was carefully examined. These children have all had Cobalt irradiation following surgery and their survival time has been excellent up to fourteen years. It is true, I have operated on another group of eight patients where there was a huge mass lesion which was solid and nothing beyond a decompression could be carried out.

SUMMARY

Infratentorial gliomas in children carry a much better prognosis than gliomas, either supratentorial or infratentorial, in adults. Investigation should be careful and treatment should be aggressive. The rewards for your efforts will be most gratifying.

Discussion

J. Hekmatpanah

The papers by DRS. WEINBERG, BUCY and HENDRICK would hardly need any specific discussion, for each is a succinct presentation of the accumulated knowledge resulting from the experience and investigations of many workers. To this, of course, is added their own personal experiences and opinions. Dr. WEINBERG presented the best that is available in neuroradiology. I wonder what DR. CUSHING would have thought of this paper had he heard it, particularly in view of his apparent concern about ventriculography which "might lead surgeons away from doing careful neurological examinations" (3).

The neuroradiologic tests, a sample of which is presented here together with the brain scan which unfortunately we did not have time to include, have been of great value in accurate diagnoses of the tumors and ultimately in a marked reduction of operative morbidity and mortality. Because of the functional localization and anatomical subdivision of the brain to frontal, parietal, occipital and temporal lobes, it became a habit to refer to the location of most cerebral tumors accordingly. But the tumors do not invade or infiltrate in respect to the above anatomical subdivisions. While focal disturbance of the function may be related to the actual location of the tumor in the related region, a good part of the signs and symptoms are falsely localized. Swelling around the tumor; obstruction of the arteries and veins and cerebrospinal fluid pathways; stretching or entrapment of the cranial nerves, in the vicinity or remote from the tumor, often cause additional signs and symptoms. Therefore, accurate localization of the brain tumor purely on "careful" neurological examination is often not possible. For reasons which will be mentioned later, approximate localization is not commonly sufficient for good operative results. Because angiography is capable of rendering histological diagnoses in meningioma, cerebral glioma, and abscesses, it is now more frequently performed than pneumography in most centers. These histological diagnoses can often be made with a high degree of confidence in meningioma and glioblastoma, but with less confidence in low grade astrocytoma and brain abscess. Sometimes the wall of an abscess is outlined well by a very fine vascular stain not ordinarily seen in neoplastic tumors. Another reason for the decline in the number of pneumographies is probably the availability of the brain scan in the last decade. The increased isotope uptake is very helpful in accurate localization of a part or all of the tumor. Although some attempt is always being made for a histological diagnosis on brain scan, the accuracy is far less than that of the angiogram. It is true that increased uptake on the brain scan is often ring-shaped for gioblastoma and is dense and adjacent to the calvarium for meningioma. But similar uptakes can be seen in brain abscess for the former, and in metastases and glioblastoma for the latter. A low-grade astrocytoma might be associated with a normal brain scan. Therefore in a patient with clinical signs and symptoms of brain tumor, in whom angiogram and pneumogram show a slight shift of the midline or ventricular system without any specific demonstration of a tumor, we take this normal brain scan in favor of a diagnosis for astrocytoma rather than against it. Finally, in so-called astrocytoma diffusium all these tests might be normal, at least in the early stages.

The computerized axial transverse tomography (CT) presented by DR. HUCKMAN seems an important recent addition to the neuroradiologic evaluation in brain tumors as a whole. Since the cerebrospinal fluid and brain tissues have a different numerical absorption value than that of brain tumors, the relation of the tumor with the ventricular system can also be evaluated. Also there seems to be some evidence that we might be able to determine the histology. This aspect, however, cannot be proven definitely at this stage and with these few cases. While a new technique, there seems to be very little doubt of its value. Nevertheless neither brain scan nor CT has so far shown the potential for being able to eliminate the use of angiography. If this were possible their value would be tremendous. The angiography, in spite of its refined technique, still inherits some complications under its use, and is also somewhat disturbing to the patient.

Sometimes the importance of preoperative histological diagnosis or impression is not fully appreciated by those who are not neurosurgeons. The design of a craniotomy flap is one of the most important aspects in the removal of a tumor. Too large or too small a flap, and also its inaccurate design, could contribute a great deal to the operative morbidity and mortality. For example, a subcortical glioblastoma beneath the sensory-motor cortex could be even widely excised without uncovering this region. The tumor could be excised through a flap anterior or posterior to the rolandic fissure, thereby decreasing the unnecessary potential damage to the sensory-motor cortex by its exposure. A similar opening for a meningioma in the same location is deleterious. During the discussion period DR. ZIMMERMAN expressed his wish "to caution" about histological diagnosis on the basis of angiography by enumerating several instances in which, in his experience, neuroradiological diagnoses have been erroneous. This of course is well understood by the neurological surgeons and is always taken into consideration. Nevertheless, an accurate preoperative localization and histological diagnosis has a lot to do with the ultimate results of most cases and it seems imperative to make it if at all possible.

DR. STEWARD's work on "Heavy Ion (Proton) Radiography in Medical Diagnosis" is still in its embryonic stage. While its potential can be appreciated, a comment on its value as compared with that of other techniques might be premature.

Many thanks to DR. BUCY for sharing with us his professional life-long experience with cerebral gliomas. His optimistic attitude toward surgical removal of glioblastomas is indeed encouraging. This optimism is certainly within the realm of technical possibility. To do the best that can be done is his philosophy, if I understand him well. Nevertheless he did not fail to convey as loudly and as clearly as he could his "incompetence" in the effective treatment of cerebral glioma. At the University of Chicago our management of glioblastoma has been essentially the same as that outlined by DR. BUCY. In my experience, however, the useful length of life gained through the operation followed by radiation was seldom more than six to nine months in most cases.

Our approach to the posterior fossa tumor has been essentially the same as is outlined by DR. HENDRICK. The preoperative shunting of posterior fossa tumors to the atrium with the purpose of remaining years after the operation has been very valuable, in our opinion, for management of posterior fossa tumors (2). We have used this procedure in almost all posterior fossa tumors with increased intracranial pressure since 1963. Similar satisfaction is also expressed by CHANDY in his "preliminary report" (1).

DR. HENDRICK prefers the prone position for operation. Sitting position, somewhat similar to Fowler's position is, in my experience, less strenuous for the patient. The chest can expand better, and the venous pressure is lower with less tendency for bleeding. Also the patient's tracheal tube and chest can be watched more easily by the anesthesiologist. In more than 100 patients for posterior fossa and upper neck operations we have not had any evidence of air embolism, using the above position.

A more vigorous approach to a more definite diagnosis for what at first could have been considered to be glioma of the brain stem but proved to be "large cysts" in the pons in five patients is commended. I thank DR. HENDRICK for bringing this to our attention.

REFERENCES

1. ABRAHAM, J., CHANDY, J.: Ventriculo-atrial shunt in the management of posterior-fossa tumors. Preliminary report. J. Neurosurg. 20, 252 (1963).
2. HEKMATPANAH, J., MULLAN, S.: Ventriculo-caval shunt in the management of posterior-fossa tumors. J. Neurosurg. 26, 609 (1967).
3. THOMSON, E. H.: Harvey Cushing, Surgeon, Author, Artist. New York: Henry Schuman 1950.

Thoughts on the Biology and Therapy of Malignant Gliomas

W. M. Kirsch, J. J. van Buskirk , D. Schulz, and K. Tabuchi

Further specifying SZENT-GYORGI's definition of cancer (1), a malignant glioma can be considered as a tumor mass composed of glial cells manifesting disordered proliferation. We remain ignorant as to what composes and comprises order in either a normal or malignant glial cell. Furthermore, great uncertainty surrounds the molecular mechanisms controlling the growth and replication of malignant glial cells, and whether cell division in normal glia is identical to mitotic events in better studied systems. If true, as WATSON has stated, that ability to effectively treat a disease depends upon our ability to understand the perturbation of molecules underlying the disorder, it is no surprise that the past record of accomplishment with malignant gliomas leaves a great deal to be desired (2). Despite this fundamental lack in basic scientific information, there are certain undeniable signals that real progress is about to be recorded in the management of malignant gliomas using an empirical approach. In a cogent analysis of predictive factors for effective chemotherapy of human cancer, SKIPPER (3) and ZUBROD (4) have shown that two objective parameters serve to herald therapeutic success. These signals, either in humans or experimental animals, consist of an increased incidence of remissions as well as a longer duration of remission. It is most likely that we can expect no dramatic or sudden breakthroughs, but must accept the inevitability of gradual progress in planned, persistent drug testing, armed with a better understanding of both cell kinetics and therapeutic agent scheduling as exemplified by the recent work of TATOR and WASSENAAR (5). Excellent experimental models for cerebral gliomas are now available, and these systems have provided evidence for tumor control, but not cure, with various drug protocols.

It is instructive to examine the reasons why malignant gliomas are so difficult to eradicate and so refractory to conventional surgical, drug and irradiation schedules. Information from a variety of studies suggests that the following factors are important and certainly merit critical examination.

i. Normal and Malignant Glial Cell Cycle Kinetics
ii. Glioblastoma Tissue Architecture
iii. Peculiarities of the Environment Surrounding a Malignant Glioma

i. Normal and Malignant Glial Cell Cycle Kinetics. There are significant gaps in information regarding the mechanism of human and malignant astrocyte proliferation. It is not certain whether classically outlined events of mitosis (as determined in less differentiated cells such as fibroblasts) actually occur during the replication of normal astrocytes. In our own laboratories, applying techniques developed by PUCK (6) we have found that the total time for completion of a malignant glial cell life cycle in culture varies from 36-48 hours. After application of colchicine blockade for 96 hours, the accumulation of cells blocked in mitosis is maximally 12% rather than the anticipated 100%. There is no evidence for decay of mitotic figures. These results are in agreement with other studies, and certainly suggest that colchicine methodology may not be applicable to life cycle analysis of either normal or malignant glial cells in culture. A significant number of glial cells may replicate by a phenomenon known as "amitosis." The latter refers to a state in which mitotic spindle cannot be detected and the nucleus actually divides by observable dissection along an elliptical plane (7). Time lapse cinematography of both normal and malignant glial cells in culture reveals mitosis with a spindle apparatus to be far more frequent in the neoplastic systems. The question of associated mesenchymal proliferation, either as a contaminant or true part of the neoplastic picture, has never been adequately resolved in these cultures (8). We too, have observed infrequent but definite examples of true mitosis in cultures of human malignant gliomas. Another unusual aspect of the cell cycle in malignant glial cells is an exceptionally high proportion of malignant glial cells, both in vivo and in vitro, residing in G_1 phase. This biological fact has important implications for the apparent lack of effectiveness of conventional carcinostatic and carcinocidal agents, as well as the peculiar problem of brain tumors where mass effects are lethal, regardless of whether tissue is replicating or not. Thus, molecular information dealing with the control of proliferation processes in glial cells, particularly pharmacologic agents that might shift a significant cohort of cells from one phase of the cell cycle to another, has important implications for the design of more effective drug and irradiation therapy. Experiments in our own laboratories with an interesting compound, racemic sodium warfarin, have provided some tantalizing clues meriting further investigation. Sodium warfarin has been known for many years to alter the natural history of both human and experimental cancer (9). The mechanism by which warfarin modifies behavior of neoplastic tissue is totally uncertain. Certain provisional observations are highly suggestive of direct effects upon the cell cycle of malignant glial cells (10). Figure 1 shows the effects of varying concentrations of the L optical enantiomer of sodium warfarin on the growth rate of human glioblastoma cells. It is apparent that even with exceptionally low

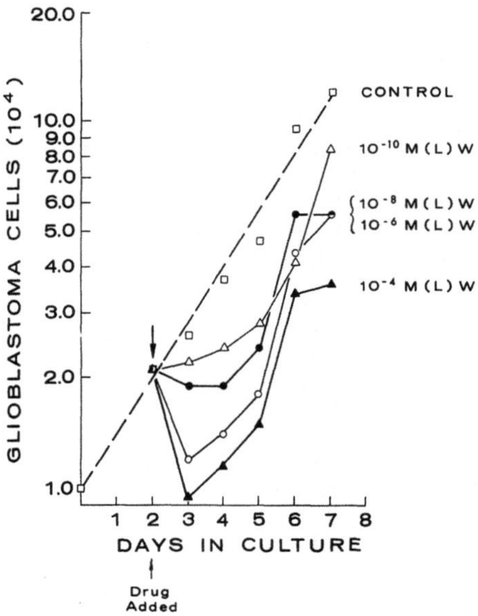

Fig. 1. Effects of varying concentrations of L-warfarin on growth and replication of human glioblastoma cells in culture. Each point represents at least three tissue culture plates counted independently

concentrations of sodium warfarin, the population cohort of glioblastoma cells remains static for approximately 72 hours to be followed by a sudden burst of cell replication over a 24 hour period, with further evidence for deceleration of replication in the next 24-48 hours. These in vitro experiments strongly suggest that L-sodium warfarin is exerting a synchronizing effect upon these cells in culture. A similar effect has been found with both the D-isomer and racemic mixture of sodium warfarin. We have turned our attention to the effect of this drug upon other cultured cells, to include six separate human glioblastomas, the murine glioblastoma, normal human embryonic glia, metastatic human bronchogenic squamous carcinoma, Hela cells and mouse-L cells, as well as the human retinoblastoma. Mouse-L cells and Hela cells have been found to be insensitive to sodium warfarin, human glial cells display moderate sensitivity, and human malignant glial cells appear to be the most sensitive. Concentrations as low as 10^{-10} molar of the L-sodium warfarin result in clear-cut suspension of cell replication for a 48 hour period. We have been able to discern a deceleration of malignant human astrocyte replication in tissue culture with exposure to warfarin at levels as low as 10^{-14} molar, or a drug concentration in the order of a few thousand molecules per cell. The plasma concentration of warfarin in patients receiving standard anticoagulation dosage ranges in the neighborhood of 10^{-5} molar, and in studies of warfarin distribution in experimental animals, tumor tissue concentrations of warfarin reach 10^{-6} to 10^{-7} molar with standard anticoagulation dosage.

Direct cell enumeration permits the straightforward calculation of statistical indices of synchrony. Asynchronous or log phase cells divide randomly at a constant rate whereas cells which are synchronized during the cell cycle will proceed through the cycle until the division point is reached. If cells are in perfect synchrony they will then all divide within a time essentially no greater than that of the division process itself. In actual practice the time span encompassing the replication of a cohort of cells, even in a well synchronized population is always greater than this hypothetical period. In multiple experiments we have obtained indices of synchrony ranging from 40-60%, utilizing the BLUMENTHAL and ZAHLER index where a value of 50% indicates good evidence for relative synchronization (11). Mechanisms accounting for these effects have been the subject of intensive investigation in our laboratory. One important factor responsible for apparent synchronization is significant detachment of neoplastic glial cells from plastic culture dishes within a few hours of exposure to sodium warfarin. The molecular derangement responsible for this detachment, encompassing 20-30% of the cells within 30 minutes, is presently unknown. It appears that a protein in the culture medium is involved, since essentially all of the administered warfarin is instantaneously protein-bound. Warfarin is bound principally to the beta globulins, and virtually not at all to albumin of fetal calf serum. The latter is used routinely in our tissue culture media. This binding is in sharp contrast to human serum in which albumin binds most of the warfarin. Provisional experiments indicate that certain membrane proteins of cultured malignant glial cells have a 30 fold greater warfarin-binding capacity than human serum albumin, raising the suspicion that this protein may serve as a cytoplasmic control factor for division.

Experimental attempts to dissect this mechanism and verify synchronization have been pursued by a variety of independent techniques. These include cytological criteria, identification of DNA synthesis (S phase) using labeled thymidine incorporation, as well as time-lapse cinemicrography. The advantage of the latter approach is that the investigator can not only alter the time scale for the surveillance of certain biological events, particularly to document the critical event of mitosis, but also examine questions pertaining to changes in cell size and detachment. Time-lapse studies of warfarin-exposed glioblastoma cells at varied concentrations demonstrate remarkable morphologic effects. Malignant glial cell motility as well as mitotic activity are markedly reduced by exposure to both D and L sodium warfarin. Furthermore, highly refractile granules have been observed to condense around the cell nucleus, with the nucleus itself becoming larger, retaining an interphase appearance. The composition of these granules has been ascertained by both light histochemical and electron microscopic studies. Samples of glioblastoma cells exposed to warfarin in vitro have been fixed for electron microscopy and these perinuclear granules have been identified as both mitochondria and neutral fat. Oil red O stains confirm that larger droplets are neutral fat. This observed clustering of mitochondria about the nucleus is identical to the intracellular morphology of cells in late S or G-2 phase prior to mitosis.

Other techniques have been utilized to ascertain the effect of warfarin on human glioblastoma cells in culture. Precision flow microfluorimetry performed by Dr. HARRY CRISSMAN at the Los Alamos Laboratories has provided a cell cycle distribution analysis to determine effects of warfarin on cell cycle progression. This technique measures the DNA content in single cells of a large population and is ideally suited for the analysis of drug effects upon cells in tissue culture. The method gives a profile of cell distributions in various phases of the cell cycle to include G_1, S, G_2 and M. After exposure to known concentrations of sodium warfarin, these provisional experiments suggest, but have not proven, that warfarin induces an accumulation of cells in S-phase. It should be emphasized that these experiments have been inconstant and subject to variation because of the phenomenon of cell detachment. By establishing a system for analyzing both floating, as well as detached cells, this highly sensitive technique will provide more and perhaps definitive information.

These in vitro experiments of drug effects have pitfalls that the investigator must be aware of. There are critical serum factors responsible for synchronization events in vitro. If the initial culture medium is replaced in three days, partial synchrony of plated glioblastoma cells can be obtained as measured by tritiated thymidine incorporation within 18-20 hours after drug exposure. This synchronization is artifactual and independent of the presence of warfarin at two concentrations, 10^{-5} M and 10^{-6} M. Glioblastoma cell synchronization is highly reproducible, but does not occur when the initial medium is only two days old and is then changed. Effects observed in vitro must be interpreted with caution in view of these observations. Synchronized fractions of cells are conceivably selected on the basis of a requirement for an unknown component of tissue culture media presumably depleted after three days growth of malignant glial cells. A special status for these cells is implied, perhaps a transition point in G_1. Multiple experiments have demonstrated however, that glioblastoma cells in vitro exposed to 10^{-6} M racemic warfarin undergo an inhibition of RNA synthesis for approximately 25 hours, and by 30 hours the inhibition of RNA synthesis is terminated. This time of inhibition coincides with late G_1 and S-phases of the synchronized fraction of cells and suggests that warfarin selectively affects cells in late G_1 and S. Conceivably, degradation of the compound prevents a similar inhibition of the daughter cells. Furthermore, plated glioblastoma cells have ^{14}C warfarin bound to a portion of their membranous proteins and as alluded to above, these proteins may be critical for cell replication.

The murine hepatoma has been studied after exposure to warfarin therapy (12). In this system, 40-50% decreases in cytoplasmic ribosomal RNA are observed specifically in the hepatoma, but not in the liver, after 4-5 daily warfarin injections. Similar treatments greatly depress rates of *in vivo* ^{14}C-orotate incorporation into hepatoma ribosomal RNA without affecting labeling of mature 18-S or 28-S RNA in the nucleus. A possible explanation of the mechanism of action of warfarin might be that it interferes with the conformation of a critical binding protein essential for progression through the cycle. The therapeutic implications of these promising observations are obvious. A relatively simple pharmacologic manipulation such as warfarin exposure, capable of synchronizing a significant population of neoplastic cells, could conceivably enhance the effectiveness of other known carcinostatic agents. For example, wide variations in radiosensitivity are detectable during different phases of the life cycle of plants, bacteria, insects, and mammalian cells. Most reports indicate that mammalian cells are most sensitive to radiation during mitosis and periods of G_1 and G_2 phases nearest mitosis, at least when chromosomal damage or cell survival are used to evaluate radiation injury (20). Another carcinostatic agent, 5-fluorouracil (5-FU), also exerts its pharmacologic effect primarily during S phase. Furthermore, though warfarin is firmly bound to human serum albumin, provisional studies reveal that the compound is even more tenaciously bound to a membrane-associated glioblastoma protein. Warfarin as conventionally administered has a very small volume of distribution within the organism, and appears to be preferentially concentrated into rat experimental glioblastomas. Experiments relating to these observations are currently in progress. These pharmacologic factors may provide important clues for the design of a more effective drug and irradiation therapy program for malignant glial tumors.

ii. Glioblastoma Tissue Architecture. Extending cell cycle experiments from *in vitro* to *in vivo* systems involves a totally new order of complexity and difficulty. What is necessary, particularly with regard to solid tumors such as the glioblastoma, is a better understanding of the cell cycle, its relationship to tissue architecture, and environmental factors such as blood supply and oxygenation. There is no question that gradients of oxygen and metabolite diffusion in solid tumors are poorly understood. Only by translating the concepts of the pathologist into biochemical terms will these problems be resolved. Our laboratory has examined the experimental subcutaneous glioblastoma in histochemical terms using the quantitative cytochemical methodology devised by LOWRY (13). Recognizing the diffusion coefficient of oxygen, and knowing the proximity of tissue cords to blood vessels, it is readily appreciated that cells within 200 microns of a capillary supply will exhibit different growth characteristics, demonstrate different drug and radiation sensitivities, and, in essence, be entirely different in respect to cell cycle characteristics from more remotely disposed cells. The bioenergetics of isolated and crowded neoplastic cells situated over 200 microns from a blood supply is mediated primarily by anaerobic means and is dependent upon different energy sources and salvage pathways. Given this cytologic variance, it is not surprising that solid tumors such as the glioblastoma appear to be more sensitive to non-cycle specific agents such as the nitrosoureas than to S-phase-specific agents such as cytosine arabinoside. This differential sensitivity is readily explainable in terms of the duration and exposure of cells to a given drug. The important practical issue is that glioblastoma cellular heterogeneity imposes serious technical difficulties in the development of *in vivo* tumor models for studying the effects of synchronization. These studies of glioblastoma tissue architecture and composition have led to the conclusion that agents with high lipid solubility, a low degree of protein binding, and non-cycle-specificity, are optimal for drug delivery into malignant brain tumors. Malignant gliomas do have an abnormal permeability to albumin, presumably because of deficient capillary endothelial junctions (14). Normal brain capillaries have a tight pentalaminar configuration whereas the thin walled vascular channels of malignant gliomas show widely patent endothelial junctions devoid of the pentalaminar structure. As a result of this capillary defect, there is open communication from the vessel to the basement membrane of the neoplastic tissue, facilitating diffusion of substances such as radio-iodinated serum albumin (RISA). RISA is preferentially concentrated into malignant gliomas, both human and experimental (15). The prediction that a biologically active drug such as warfarin is preferentially concentrated into a malignant brain tumor on the basis of binding to a carrier protein has been fulfilled experimentally in our laboratory.

iii. Peculiarities of the Environment Surrounding a Malignant Glioma. A third critical factor which must be considered in the design of therapy, and about which we have only meager understanding, yet some empirical control, is the relationship of the malignant glioma to normal brain. The one pharmacologic manipulation appearing to have altered the morbidity and mortality of malignant gliomas, at least in the early stages of diagnosis or operative intervention, has been the introduction of steroid therapy for the control of cerebral edema (16). Adjunct steroid therapy has been widely accepted clinically and is clearly beneficial to patients by reducing associated cerebral edema. In addition to malignant glial cell necrosis secondary to drugs and irradiation, considerable cell attrition occurs normally, and presumed release of lysosomal enzymes results in secondary cerebral edema or brain swelling. In our clinical experience, we have found that an effective combination of drug and irradiation therapy on malignant gliomas will result in an aggravation of clinical symptomatology necessitating emergency operative decompression. Large quantities of necrotic cellular debris can be removed from the tumor bed with temporary clinical improvement of the patient. SANO provides indwelling intracavitary tumor drainage in patients receiving combined radiosensitization and irradiation therapy to prevent this worsening of clinical symptomatology associated with tumor cell death (17). Thus, disposition of dead and necrotic tissue from a treated malignant glioma is an exceptionally important factor governing the success or failure of chemotherapy. These relationships are to be contrasted with treatment of solid tumors elsewhere in the body where cell disposition can be accomplished without the hazards of increased intracranial necrotic pressure.

Summarizing our own clinical experiences over the past six years, we have treated over 120 patients with a variety of surgically incurable tumors of the neuraxis. The treatment program has consisted of anticoagulation with sodium warfarin, irradiation therapy, and carcinostatic agents to include 5-fluorouracil, cytosine arabinoside, and CCNU. Certain neoplasms have responded favorably to this treatment program. These studies have not been controlled clinical trials, but rather a Phase II type study in which the effects on our patients are compared to previous literature survival values or to effects on other patients hospitalized at our institution and treated by other methods. The median survivorship of a patient with a documented glioblastoma treated with warfarin, 5-FU, and ionizing radiation is approximately 11 months. Two patients have survived for 2.5 years with glioblastomas involving the thalamus. Our clinical impression is that the quality of survival has been worth the effort and that drug toxicity has been minimal. Ten patients, with multiple intracranial metastases in some cases, have a median survivorship of 7.6 months in comparison to an anticipated 2 - 3 months. In these metastatic tumor cases, particularly squamous carcinoma of the bronchus, quality of survival has been remarkably good. Our most gratifying chemotherapy patients are those with tumors classified as astrocytoma grades II and III, and the brainstem glioma of childhood. Long term survival measured in years may be clearly anticipated with drug and irradiation therapy of these tumors. Interpretation of this data must be considered provisional because of the uncertainty of the natural history of these tumors with conventional radiation therapy, but it does appear that statistically significant improved survival is being offered. It must be emphasized however, that we know of no patients who have been cured of their disorders by this treatment program. We believe that the signals of increased remission as emphasized by ZUBROD and SKIPPER suggest and anticipate further progress.

Other promising therapeutic avenues for malignant gliomas deserve attention. One important approach deals with the use of naturally occurring molecules to augment further differentiation of neoplastic mammalian neural tissue. The best studied model is the mouse neuroblastoma, a neoplasm whose human counterpart for unknown reasons may undergo spontaneous differentiation and remission. PRASAD and HSIE have demonstrated that cyclic AMP inhibits the growth of the neuroblastoma cell population with attendant cytologic differentiation of these cells in culture (18). SANO is currently treating malignant gliomas by intracavitary cyclic AMP insertion and the results of these ongoing clinical trials will be of great interest. Thus, the use of naturally occurring small molecules such as cyclic AMP, and a better understanding of specific cell receptors for these substances, may play an enormous role in making highly malignant cells behave in a more orderly fashion. Recent work of CUATICO is also of great interest (19). CUATICO has isolated a reverse transciptase from malignant gliomas, suggesting that a type C RNA tumor virus is a putative agent for malignant transformation of normal glia cells through a DNA intermediate. Perhaps directing drug and radiation therapy toward DNA biosynthesis is an attack on the wrong molecule, and more attention should be given to various drugs which inhibit the specific enzyme reverse transcriptase.

Advances in radiation therapy technology are clearly a future prospect. In recent years, a variety of high energy linear transfer (LET) radiations from accelerators have become available for radiotherapeutic application. Most promising among these particles are the negatively charged π mesons, referred to as "pions." These particles have two important advantages for radiation therapy of malignant gliomas. First, there is an increase in dose (2-4 times in soft tissue) at the end of a pion's track as compared to that delivered to the surface. The tissue in fact appears to mediate this effect by the release of low energy highly ionized particles by nuclei, primarily oxygen, which fragment after capturing a slowly moving pion. Secondly, the high LET of heavy particles released at the end of the pion's track will increase the biological effectiveness of the tissue radiation, thus reducing the probability of cellular repair and decreasing the influence of the tissue concentration of oxygen. As noted above, glioblastomas are essentially anaerobic tumors except for a thin crust. Though at present no accelerator can furnish pions at the dose desired or in the time frame necessary for therapy, these technical limitations will undoubtably be surmounted in the near future.

At present, our understanding of immunologic factors governing malignant glioma therapy remains qualitative and poorly characterized. Paradoxically, current management consists of potent suppression of the patient's immune response by x-rays, steroids, and conventional chemotherapeutic agents. It seems illogical to combine these modalities with BCG therapy designed to augment the immune response. The most promising immunological lead appears to be the demonstration of circulating humoral factors (in patients with malignant gliomas) which inhibit phytohemagglutinin activation of lymphocytes (20). This work, which has been confirmed by Dr. HARRY YOUNG at Richmond, Virginia may serve as a valuable diagnostic test as well as a means for monitoring the subclinical course of malignant glial tumors. Chronic immuno-suppression in humans results in a high incidence of certain forms of malignant tumors within the neuraxis, particularly tumors of the lymphoma series (21). We have had, for example, a patient under our care who developed a primary lymphoma of the thalamus in the face of immunosuppression therapy subsequent to renal transplantation. This patient sustained a complete remission after elimination of immunosuppression therapy as well as application of radiation therapy to a neoplasm. With regard to surgical innovations for malignant gliomas, the work of SUTTON holds promise (22). SUTTON has utilized microwave heating in order to favorably alter both the blood brain barrier and glioma blood flow as determined by the distribution of peroxidase. These studies suggest that local application of heat can increase tumor blood flow and facilitate the accumulation of drugs within the neoplasm.

In summary, it can be stated that a better understanding of the fundamental biology of malignant gliomas is accruing at a significant rate. Furthermore, this newfound knowledge is being utilized to develop treatment schedules offering an increase both in the length and the frequency of remission to patients with surgically incurable tumors involving the neuraxis, primarily malignant gliomas. This palliation, modest as it may be, serves as an important indicator for the significant advances to be hopefully registered within the near future.

This research was supported by VA Grant #374-4964-01, by the Pediatric Oncology Center Grant NCI-CA-12247, and by the Adrian Hyslop Brain Tumor Research Fund.

REFERENCES

1. SZENT-GYORGYI, A.: "On Cancer and Cancer Research", On Cancer and Hormones. Chicago and London: University of Chicago Press 1962.
2. WATSON, J. D.: Molecular Biology of the Gene. Chapter 16, p.442. New York: W. A. Benjamin, Inc. 1965.
3. SKIPPER, H. E.: "Successes and Failures at the Preclinical Level; Where Now?" Seventh National Cancer Conference Proceedings 1973.
4. ZUBROD, C. G.: "The Basis for Progress in Chemotherapy," Cancer J. for Clinicians 23, 202 (1973).
5. WASSENAAR, W., TATOR, C. H., SO, W. S.: "The Development of an Intra-Cerebral Glioma Model for Brain Tumor Chemotherapy" J. Neurosurg. 39, 293 (1973).
6. PUCK, T. T.: "Life Cycle Analysis of Mammalian Cells. I. A method for localizing metabolic events within the life cyle, and its application to the action of colcemide and sub-lethal doses of X irradiation." Biophys. J. 3, 379 (1963).
7. LAPHAM, L. W.: "Cytologic and Cytochemical Studies of Neuroglia: The Study of the Problem of Amitosis and Reactive Protoplasmic Astrocytes" Am. J. Path. 41, 1 (1962).
8. HUGOSSON, R., KALLEN, B., NILSSON, O.: "Neuroglia Proliferation Studied in Tissue Culture" Acta Neuropathol. 11, 210 (1968).
9. MICHAELS, L.: "Cancer Incidence and Mortality in Patients Having Anticoagulation Therapy" Lancet 1964 II 832.
10. KIRSCH, W. M., VAN BUSKIRK, J. J., SCHULZ, D.: "Effects of Sodium Warfarin and Other Carcinostatic Agents on Malignant Cells: The Study of Drug Synergy" J. Exptl. Med. (in press).

11. BLUMENTHAL, L. K., ZAHLER, S. A.: "Index for Measurement of Synchronization of Cell Populations" Science 135, 724 (1962).
12. VAN BUSKIRK, J. J., KIRSCH, W. M.: "Loss of Hepatoma Ribosomal RNA During Warfarin Therapy" Biochem. Biophys. Res. Commun. 52, 562 (1973).
13. KIRSCH, W. M., SCHULZ, D., LEITNER, J. WAYNE: "The Effect of Prolonged Ischemia upon Regional Energy Reserves in the Experimental Glioblastoma" Cancer Res. 27, 2212 (1967).
14. LONG, D. M.: "Capillary Ultrastructure and the Blood-Brain Barrier in Human Malignant Brain Tumors" J. Neurosurg. 32, 127 (1970).
15. TATOR, C. H., SCHWARTZ, M. L.: "Autoradiography of Brain Tumors" from The Experimental Biology of Brain Tumors, Chapter 6, p.194, Eds. KIRSCH, PAOLETTI and PAOLETTI. Springfield, Ill.: Charles C. Thomas 1972.
16. FRENCH, L. A., GALICICH, J. H.: "The Use of Steroids for the Control of Cerebral Edema" Clin. Neurosurg. 10, 212 (1964).
17. TAKAKURA, K., SABO, K.: "Sites of Action of Chemotherapeutic Drugs on Brain Tumor Metabolism" from The Experimental Biology of Brain Tumors, Chapter 14, Eds. KIRSCH, PAOLETTI and PAOLETTI. Springfield, Ill.: Charles C. Thomas 1972.
18. PRASAD, K. N., HSIE, A. W.: "Morphological Differentiation of Mouse Neuroblastoma Cells Induced In Vitro by Dibutyryl Cyclic AMP" Nature New Biol. 233, 141 (1971).
19. CUATICO, W., CHO, J. R., SPIEGELMAN, S.: "Particles with RNA of High Molecular Weight and RNA-Directed DNA Polymerase in Human Brain Tumors" Proc. Nat. Acad. Sci. 70, 2789 (1973).
20. HENDEE, W. R., KIRSCH, W. M.: "Effects of Ionizing Radiation Upon Normal and Neoplastic Neural Tissue" from The Experimental Biology of Brain Tumors, p. 490, Eds., KIRSCH, PAOLETTI and PAOLETTI. Springfield, Ill.: Charles C. Thomas 1972.
21. BROOKS, W. H., NETSKY, M. G., NORMANSELL, DAVID E., HORWITZ, DAVID A.: "Depressed Cell-Mediated Immunity in Patients with Primary Intracranial Tumors" J. Exp. Med. 136, 1631 (1972).
22. PENN, I.: Malignant Tumors and Organ Transplant Recipients, Berlin. Heidelberg New York Springer 1970.
23. SUTTON: Scientific Program, Paper No. 76.Proc. Am. Assoc. Neurol. Surg. Annual Meeting, 1974.

Chemotherapy: Current Results and Future Prospects

C. B. WILSON

As chemotherapy becomes effective in an increasing number of neoplastic conditions, the future prospects for the treatment of malignant gliomas become brighter. While surgery and irradiation have increased the quality and duration of survival after treatment, these two modalities alone offer little prospect of improvement by further refinement. The addition of chemotherapy and, possibly, immunotherapy, represents a therapeutic approach deserving high priority if we intend to take full advantage of advancements made in the management of tumors outside of the CNS. My purpose here is to present the current state of brain tumor chemotherapy and to indicate what I believe will be the future direction of our therapeutic efforts.

Like other solid tumors, glioblastomas, the prototype of the group of tumors under consideration, contain three populations of cells in terms of their kinetic behavior (1). In a first compartment are actively dividing cells, i.e., cells undergoing continuous replication. In small tumors, proliferating cells predominate and account for the short doubling time of most small tumors. As tumor mass increases, the availability of nutrients becomes suboptimal, and a steadily increasing portion of cells enter a non-proliferating phase. In the second compartment non-proliferating cells retain the potential for cell division, but because of unfavorable local conditions they have left the cell cycle as a temporary expedient. As tumors become progressively larger an ever-increasing portion of cells leave the cell cycle following mitosis and enter a resting or dormant state. The availability of oxygen is believed to govern the kinetic behavior of individual cells within a tumor cell population, and in experimental tumors, at least, tumor cell proliferation exceeds the capacity of blood vessels to nourish an expanding population of cells. Whereas glucose can diffuse a considerable distance, the diffusional characteristics of oxygen limit the availability of oxygen to cells most distal to the nearest capillary. It is the inability of blood vessels to keep pace with an expanding tumor cell population that accounts for deceleration of growth rate as tumors become progressively larger. In a third compartment, sterile and dead cells accumulate as tumor growth proceeds. Sterile cells, as opposed to resting cells in the second compartment, have lost their ability to divide as a consequence of some defect in genetic material. Cells dying either because of genetic defects or because of nutritional deprivation accumulate as the rate of cell death exceeds the rate of cell loss, the latter occurring by a process of lysis and absorption. When a glioblastoma has assumed clinical proportions, less than 1/3 of histologically viable cells are in the process of active cell division, and the bulk of the remainder are in a non-proliferating phase. An unknown minority of histologically viable cells have permanently lost the capacity to divide, but since these cells will add nothing to the tumor cell population, they are of relatively little concern to the chemotherapist. Factors governing the progressive accumulation of dead cells are exceedingly complex and little known, but dead and dying cells contribute substantially to the bulk of glioblastomas. Studies by HOSHINO in our Department have defined the kinetic parameters of glioblastomas, and these are indicated in Table 1.

Table 1. Kinetic parameters of human glioblastomas

T_S (DNA synthesis)	5-10 hours
T_C (cell cycle time)	48-72 hours
LI (labelling index)	5-10%
GF (growth fraction)	30-40%

SELECTION OF DRUGS, ROUTES OF ADMINISTRATION AND DRUG SCHEDULES

Pharmacokinetic considerations argue for the use of drugs that cross the intact blood-brain barrier (BBB) to enter normal or near-normal brain adjacent to an intracerebral tumor. Admittedly, most water soluble drugs and large molecules cross defective blood vessel walls to enter the tumor's extracellular space, but because these substances are excluded from the surrounding brain by an intact BBB, the steep concentration from tumor to brain and CSF results in rapid diffusion of drug out of tumor and down a concentration gradient to brain and CSF. For drugs such as the lipid-soluble nitrosoureas, systemic administration results in similar drug entry into brain and into tumor, eliminating a concentration gradient. Consequently high concentrations of drug are attained and maintained at the tumor's growing edge where proliferative activity and viability are highest. Although water soluble compounds, such as Methotrexate, can be effectively administered intrathecally, the kinetic characteristics of most malignant gliomas and their frequent location at a distance from subarachnoid and ventricular CSF argue against primary reliance upon cell cycle specific agents administered by this route.

Oncolytic agents can be separated into two groups on the basis of vulnerable cells within a tumor cell population. Cell cycle specific (CCS) drugs affect only those cells in some phase of the cell cycle at the time of drug exposure, and non-cycling cells are unharmed. Many CCS agents affect only those cells in one phase of the cell cycle, usually the S phase of DNA synthesis. Clearly, a CCS drug is most effective against tumors containing a high proportion of proliferating cells, i.e., tumors possessing a high growth fraction[1] and relatively ineffective against a tumor composed predominantly of non-proliferating cells. By contrast, cell cycle non-specific (CCNS) drugs affect all tumor cells more or less equally irrespective of their proliferative state at the time of drug exposure. Consequently CCNS drugs should be selected for use against tumors with a significant proportion of non-cycling (non-proliferating) cells. Tumors with a high growth fraction, e.g., choriocarcinoma and most very small solid tumors, are vulnerable to the action of CCS agents such as Methothrexate. On the other hand, tumors that have a relatively low growth fraction, e.g., glioblastomas, are better

[1]growth fraction (GF) = $\frac{\text{(no. of proliferating tumor cells)}}{\text{(total no. of tumor cells)}}$.

targets for CCNS agents such as ionizing irradiation, alkylating agents and the nitrosoureas since a CCS drug would affect a minority of the total tumor cell population.

Chemotherapy, antibacterial as well as antitumor, follows first order kinetics, i.e., for any given dose, an effective drug kills a fixed proportion of cells. For example, a drug capable of 1 $\log_{10}$ (90%) kill will reduce a vulnerable population from 10^{11} to 10^{10} cells, and with smaller populations the proportional reduction is the same (10^6 to 10^5 cells). It follows logically that elimination of the last cell will require multiple doses administered over a long period of time, likely many months in the case of anaplastic gliomas.

Drug schedules are all-important, even critical, to success. Two examples will illustrate the point. Two drugs, A & B, both individually effective against an experimental tumor, when administered in single doses by different schedules can produce (1) less effect than either drug alone, (2) an additive effect, or (3) a strikingly synergistic effect. The only variables, dose remaining constant, are the interval of time separating a single dose of the two drugs and the order, i.e., A-B or B-A, of their administration. The second example relates to the interval between successive doses or courses of a drug (or drugs). The period of time required for recovery of bone marrow determines the interval between doses of most currently useful drugs. In the case of nitrosoureas this recovery period is in the range of 6 weeks. Tumor cells that survive a course of treatment immediately proliferate in an effort to re-populate the decimated tumor cell population. Unless a subsequent course can be administered before the proliferation of unharmed cells restores the tumor to its original size, succeeding courses of therapy will not achieve a progressive reduction in cell mass with final elimination of the last cell. To date drugs have been administered with little thought to dose-dependent factors and in relative ignorance of kinetic considerations such as quantitative cell kill and repopulation between drug courses. Information of this sort becomes even more important in planning multiple drug therapy.

TUMOR CELL DEATH AND DEAD CELL ELIMINATION

Following effective treatment of a tumor <u>in situ</u>, whether by radiotherapy or chemotherapy, cells are killed more or less rapidly. The original mass effect is magnified by at least two factors: first, a non-viable cell swells, perhaps as much as twice its normal size; and second, by-products of cell death lead to the accumulation of extracellular fluid in tumor and in brain adjacent to tumor. These and doubtless other factors aggravate pre-existing neurologic deficits and intensify precariously compensated intracranial hypertension. While adrenal corticosteroids favorably modify this situation, the paradoxical consequence of tumor cell kill can be uncontrollable increased intracranial pressure.

CLINICAL EXPERIENCE WITH SINGLE AND MULTIPLE AGENTS

<u>Patients Selected for Treatment</u>. All patients had identifiable tumors and were deteriorating at the time of treatment. All tumors were defined by brain scan and appropriate contrast studies, angiography prov-

ing adequate with rare exceptions. Progressive neurologic deterioration was established by history. The greatest number of patients had tumors recurring after subtotal removal and postoperative irradiation, and in these patients radiotherapy had been completed at least 3 months prior to starting chemotherapy. Patients with metastatic brain tumors were treated after histologic confirmation of the primary site. In a third category were patients with intact primary tumors, a few of which were confirmed by needle biopsy before referral. The majority harbored an unverified mass whose angiographic characteristics and location supported no other diagnosis, and either the referring neurosurgeon or the patient decided to begin chemotherapy without an histologic diagnosis. In this regard we have autopsy confirmation in 87% of all deaths, and our only diagnostic errors involved two primary reticulum cell sarcomas treated with the presumptive diagnosis of malignant gliomas.

We had one question in this study: does the tumor regress after chemotherapy? This constitutes the traditional Phase II trial seeking evidence of therapeutic activity. In the present study we excluded patients receiving immediate postoperative adjuvant chemotherapy, a situation in which survival (or time to recurrence) rather than tumor regression is the criterion of response.

Methods of Evaluation. After exploring and discarding serial angiography, we applied two indications of therapeutic effectiveness: serial neurologic examinations and serial brain scans. The type of responses are defined in Table 2. Patients failing to survive for 2 months after starting treatment were eliminated as non-evaluable. All patients surviving 2 months or longer were classified by the criteria indicated in Table 2. Evaluation preceded each course of treatment and every course was either administered or started in the hospital.

Table 2. Response to chemotherapy

Designation	Clinical condition	Brain scan
A. Responder	Improved	Improved
B. Probable responder	1. Improved	Unchanged
	2. Unchanged	Improved
	3. Unchanged (for 3 months with malignant tumors)	Unchanged
C. Non-Responder	Worse	Worse

Slightly more than half of patients required maintenance glucocorticoid therapy at some time during chemotherapy, the majority being started on glucocorticoids either before referral or before we began chemotherapy. We avoided mistaking steroid effect for chemotherapeutic response by designating as responders only those patients who improved while receiving either the same or a reduced dose of steroid. In most cases steroids were discontinued in responding patients. A retrospective analysis indicated identical response rates in patients receiving steroids and in patients not receiving steroids.

Results with Single Agents. The details of treatment can be found in earlier publications (2, 3, 4). For present purposes the results obtained with three agents will be summarized.

With 1, 3 bis-(2-chloroethyl)-1-nitrosourea (BCNU) we obtained a 47% response[2] (27/57 patients) for a median duration of 9 months. The results with 1-(2-chloroethyl)-3-cyclohexyl-1-nitrosourea (CCNU) were similar but slightly inferior: a 44% response[2] (16/36 patients) for a median duration of 6 months. Both nitrosoureas rapidly cross the normal blood-brain barrier and both produce delayed depression of bone marrow.

The third effective single agent, N-isopropyl-OC-(2-methylhydrazine)-p-toluamide hydrochloride (Procarbazine), also enters normal brain and CSF. Like CCNU, procarbazine has the advantage of oral administration. Procarbazine proved to be an effective drug: a 52% response[2] (14/27) for a median duration of 6 months.

Results with Multiple Agents. Our first multiple drug protocol involved the simultaneous administration of Procarbazine, CCNU, and Vincristine. We selected Vincristine because it has shown activity against astrocytomas and medulloblastoma and because it has little depressant effect on the bone marrow. Vincristine's neurotoxicity is readily detected and usually reversible if treatment is stopped.

In courses repeated at intervals of 4 weeks, the three drugs were administered as follows: Procarbazine, 100 mg/m^2 days 1-14; CCNU, 75 mg/m^2 on day 1; and Vincristine, 1.4 mg/m^2 on days 1 and 8. This combination achieved a 57% response[2] (25/44 patients). Because a number of patients continue to respond we have not calculated the median duration of response. Particularly impressive have been responses in 5 of 6 patients harboring recurrent medulloblastomas, although in all cases treatment was comprised because of inadequate bone marrow reserve as a consequence of previous spinal irradiation.

A second multiple drug protocol combined CCNU, Vincristine and cyclophosphamide (Cytoxan). After observing only one brief response in 14 evaluable patients we abandoned this combination because of its ineffectiveness.

FUTURE PROSPECTS FOR CHEMOTHERAPY

The potential of chemotherapy is illustrated by two patients who were first treated five years ago. One, an adult, presented with an astrocytoma recurring after surgery and irradiation. The other, a boy of fourteen, was referred with spinal, posterior fossa and supratentorial tumors one year following operation and radiotherapy for a fourth ventricular ependymoma. Both patients received BCNU for a total period of two years, and they are now alive without evidence of disease 30 and 36 months later. While these two responses are exceptional, they indicate that chemotherapy, under the best circumstances, can be highly effective.

While chemotherapy using a single drug has achieved dramatic results in a few solid tumors, notably choriocarcinoma and Burkitts lymphoma, a single agent approach to the majority of solid tumors has few advocates. Combination chemotherapy has been highly effective against several tumors, and the greatly improved results have encouraged further laboratory and clinical research involving multiple drugs administered

---

[2] includes unequivocal and probable responders.

concurrently and sequentially. In the laboratory we are defining an optimal schedule for the combination of BCNU and Procarbazine, a two drug combination that is highly promising in a current clinical trial.

Multimodality therapy has been applied to a number of non-neural tumors, e.g., surgery, radiotherapy, chemotherapy, and now immunotherapy in the treatment of breast cancer. With primary brain tumors, distant micrometastatic foci do not require separate consideration, and the intracranial mass of approximately 100 grams (10^{11} cells) can receive the therapist's undiverted attention. Viewed in quantitative terms, a 100 gm. mass must be eliminated to effect a cure. Assuming that immune mechanisms (possibly enhanced by specific or non-specific immunotherapy) can eliminate a tumor reduced to 10,000 (10^4 or 4 logs) cells, surgery, radiotherapy and chemotherapy must achieve a 7 $\log_{10}$ reduction of the tumor cell population. Under the best of circumstances the surgeon can remove 90-99% of a glioma, accomplishing a 1 or 2 $\log_{10}$ reduction and leaving 10 logs (10 gm.) or 9 logs (1 gm.) of residual tumor. Optimal and concurrent application of radiotherapy and combination chemotherapy have the potential of eliminating an additional 5 or 6 logs and reducing the tumor cell population to a number (4 logs) that in theory can be handled by the host's enhanced immune mechanisms.

Our current experiments with multimodality therapy in animal models indicate that the timing of chemotherapy in respect to surgery and radiotherapy is critical. If we can combine these modalities and eliminate large, symptomatic gliomas in rats, the therapeutic plan can be cautiously applied to man. In the forseeable future we consider this the most reasonable approach to the treatment of human gliomas. We are not ignoring the potential value of immune enhancement, but we have little reason to anticipate a major role for immunotherapy during the next few years.

REFERENCES

1. HOSHINO, T., BARKER, M., WILSON, C. B., BOLDREY, E. B., FEWER, D.: Cell kinetics of human gliomas. J. Neurosurg. 37, 15 (1972).
2. FEWER, D., WILSON, C. B., BOLDREY, E. B., ENOT, K. J.: A phase II study of 1-(2-chloroethyl)-3-cyclohexyl-1-nitrosourea (CCNU). Cancer Chemother. Rep. Part I, 56, 421 (1972).
3. FEWER, D., WILSON, C. B., BOLDREY, E. B., ENOT, K. J., POWELL, M. R.: The chemotherapy of brain tumors. Clinical experience with carmustine (BCNU) and vincristine. JAMA 222, 549 (1972).
4. KUMAR, A. R. V., RENAUDIN, J., WILSON, C. B., BOLDREY, E. B., ENOT, K. J.: Phase II study of N-Isopropyl-OC-(2-Methylhydrazine)-p-toluamide hydrochloride (Matulane, Procarbazine HC1: NSC-77213) in the treatment of brain tumors. J. Neurosurg. 40, 365 (1974).

Conventional Radiation Therapy of Gliomas*

G. E. Sheline

Assessment of the value of radiation therapy in treatment of the gliomas is difficult. Neurosurgeons and radiation oncologists have had individual opinions regarding the usefulness of radiation therapy and have either given or withheld irradiation accordingly. Until a recent, and as yet unpublished, study of malignant gliomas (1) was initiated, these opinions had never been subjected to the rigors of prospective randomized investigation. Many retrospective analyses have been reported but often details of the surgical and radiation treatments have been omitted. In some reports it is even impossible to ascertain which patients received irradiation. Further compounding the problem is the fact that the natural history, resectability, and responsiveness to irradiation of the various gliomas differ widely.

For analysis of results of treatment gliomas must be divided according to histologic classification and comparison based on common definitions. The location and extent of the lesion should also be considered. The necessity for numerous subdivisions means that, even in large treatment centers, many years are required to collect an adequate number of patients for study. Furthermore, during such periods of time surgical and irradiation techniques change.

This presentation relies largely on our experience at the University of California School of Medicine at San Francisco (UCSF), but is supplemented with selected data from the literature. To avoid repetition the treatment policy for gliomas which has been practiced at UCSF will be stated at the outset. Except in the case of tumors inaccessible by virtue of location, all patients suspected of having a brain tumor were surgically explored, at which time the lesion was biopsied and removal usually attempted. In most instances surgical resection was conservative with an effort made not to increase materially the existing neurologic deficit. Depending upon the opinion of the neurosurgeon involved, the patient was then referred for radiation therapy. No patient was accepted for irradiation unless surgical removal of the tumor was thought to be incomplete. Over the years radiation therapy techniques have varied. For the group of tumors under consideration, the principle has been to irradiate the primary tumor site using fields large enough to allow for uncertainty of tumor extent and doses approaching the suspected tolerance of the central nervous system (CNS). In none of the UCSF cases reported here did the fields include the whole brain or the spinal cord. In recent years the usual dose has been 5000 to 5500 rads delivered in 5 fractions per week with about 180 rads per fraction.

*Presented at the Symposium on Gliomas, sponsored by the University of Chicago, Chicago, Illinois, on May 19, 1974.

ASTROCYTOMA

Several years ago a review of the results of surgical treatment and treatment by irradiation of intracranial neoplasms was instituted at UCSF. With respect to astrocytomas, the review has been completed only for children. These results for children will be supplemented by the data reported by BOUCHARD (2) for patients of all ages.

Table 1. Astrocytoma survival rates

Series	Numbers of patients	3 yr	Survival 5 yr	10 yr
UCSF (age ≤ 20 yr)				
surgery + irradiation	21	-[a]	85% (18/21)	74% (14/19)
MARSA et al. (4) (age < 15 yr)				
surgery + irradiation	15	-	72%	-
LEVY and ELVIDGE[b]				
surgery only	42	52%	26%	-
surgery + irradiation	45	62%	36%	-
BOUCHARD (2)				
surgery + irradiation	105	64%	49%	-

[a] - = not reported.
[b] As cited by J. BOUCHARD (2).

Between the middle 1940's and 1968, 21 children with incompletely resected astrocytomas received radiation therapy at UCSF (Table 1) (3). Fifteen had cerebellar astrocytomas described at surgical operation as having invaded pons, brain stem, vermis, or cerebellar peduncles. The 5-year and 10-year recurrence-free survival rates were 85% and 74%, respectively. Five patients have survived more than 20 years. Regarding the quality of survival, 10 of the present 16 survivors are neurologically normal or very nearly so. The neurologic status of one patient is unknown. The other 5 have deficits ranging from aphasia to hemiplegia or mental retardation. In 4 of these patients the neurologic damage was due to the tumor; the fifth patient underwent 2 courses of radiation therapy and 2 surgical procedures and it is probable that his mental retardation was treatment-induced.

Although the natural history of astrocytomas is relatively long, the finding of a 74% recurrence-free survival rate at 10 years suggests that radiation has a lasting beneficial effect. These findings are supported in a recent study of MARSA et al. (4), who reported a 72% 5-year survival rate for a group of 15 children with irradiated gliomas.

BOUCHARD (2) reviewed the results of LEVY and ELVIDGE and combined them with his own data for patients of all ages with astrocytomas. LEVY and ELVIDGE has a 5-year survival rate of 36% when surgical treatment was followed by irradiation, compared with 26% for nonirradiated patients. BOUCHARD, using surgical treatment and irradiation, reported a 5-year survival rate of 49%. These data suggest irradiation increased the survival rate, but the numbers of patients are small, the series are nonrandomized, and the interpretation of differences is open to question.

To conclude, radiation therapy of astrocytomas does lead to improvement in certain patients and the benefit may persist for many years. Furthermore, it is probable that some patients are cured. Whether such responses occur with sufficient frequency that the effect of irradiation on the 5-year survival rate can be demonstrated awaits a controlled study.

MALIGNANT GLIOMA AND GLIOBLASTOMA MULTIFORME

Differing concepts regarding classification of the malignant glial tumors has led to confusion. One school of thought divides these lesions into grade III or IV astrocytoma; another classifies them as malignant glioma, malignant astrocytoma, poorly differentiated astrocytoma, or glioblastoma multiforme. Unfortunately, in discussions of results of treatment, these lesions have often been lumped into a single group.

Differences in classification of the malignant gliomas may explain the differences of opinion regarding the efficacy of irradiation. Many neurosurgeons, at least in the past, thought irradiation useless for glioblastoma. On the other hand, BOUCHARD (2) reported a substantial increase in survival time in a 7% 5-year disease-free survival rate for patients irradiated for glioblastoma multiforme. It is, however, not clear what BOUCHARD included under this diagnosis. TAVERAS, THOMPSON, and POOL (5) reported a 4% 5-year survival rate for irradiated glioblastomas, but this is known to have included 10% of grade III astrocytomas. KRAMER (6) classifies malignant glial tumors as astrocytoma grade III or IV; in irradiated patients he reports a 22% (5/23) 5-year survival rate for grade III and no survivors for grade IV lesions.

Table 2. Malignant glioma and glioblastoma multiforme (UCSF Series)

Treatment	Malignant glioma	Glioblastoma Multiforme
Not irradiated	10 (all DOD[a])	50 (all DOD)
Irradiated	39	40
M/F	21/18	31/9
age at diagnosis		
0-20	8	2
21-40	12	3
41-60	14	31
61	5	4
DOD (median survival)	31 (10 mo)	40 (10 mo)
5-year survival	7/39 (18%)	None
NED (yr)[b]	5 (5,6,8,9,10)	None

[a]Died of the disease.
[b]No evidence of disease (years postoperative).

Patients in our series are classified into 2 groups (Table 2). Group I refers to those patients having malignant gliomas pathologically diagnosed as malignant astrocytoma, poorly differentiated astrocytoma, malignant glioma, or astrocytoma grade III. The number of patients is too small to permit a further breakdown of this group. Group II includes patients with lesions diagnosed as glioblastoma multiforme. Of 139 patients, 49 had malignant glioma (Group I) and 90 glioblastoma multiforme (Group II). Ten patients with malignant glioma and 50 with glioblastoma multiforme received no radiation therapy and all are dead of the glioma. Nine of the 10 patients with nonirradiated malignant glioma did not survive long enough to undergo radiation therapy. The median survival for the 50 patients with nonirradiated glioblastoma multiforme was less than one month, and only 3 of these patients survived beyond 3 months. Thus, because of their short survival time, the nonirradiated patients represent a different mix of patients than those who were irradiated and they cannot be used as a control group for the irradiated patients.

Turning to the 39 patients with irradiated malignant glioma and the 40 with irradiated glioblastoma multiforme, we see differences between the 2 groups. The male to female ratio was approximately 1 to 1 for patients with malignant glioma but 3 to 1 for those having glioblastoma multiforme. Half of the patients with malignant glioma were under 40 years old, whereas only 12% of those with glioblastoma multiforme fell into this age group. The 5-year survival rate for malignant glioma was 18% versus 0% for glioblastoma. The median survival rate for those dying of the disease was 10 months for each group. Of the 7 patients with malignant glioma who survived 5 years, 4 are still alive and leading productive, essentially normal, lives. One died of intercurrent disease at 8 years. One died of recurrence between 6 and 7 years after treatment. The seventh patient lived 12 years, although institutionalized during the latter half of this period. Whether her disability and death were due to the treatment or resulted from the glioma is unknown; necropsy was not permitted.

Whether or not our classification of aggressive gliomas is optimal, it appears that, when reporting results, some such subdivisions should be used. Our patients with glioblastoma multiforme, like those of KRAMER (6) with astrocytoma grade IV, all died of the disease. On the other hand, the irradiated patients with malignant glioma had an 18% 5-year recurrence-free survival rate. This may be compared with KRAMER's 22% for astrocytoma grade III.

EPENDYMOMA

Ependymomas, neoplasms derived from ependymal cells, are generally considered among the most radioresponsive of the gliomas. While a few may be totally resectable, particularly those in the fourth ventricle that arise tongue-like from a narrow base, the majority cannot be totally resected and require radiation therapy. Most ependymomas are slow-growing, circumscribed neoplasms which seldom spread and which can be treated locally. Some disseminate through the cerebral spinal axis in a manner similar to that of the medulloblastoma. When considering treatment of the ependymomas two questions arise. First, is the incidence of dissemination sufficient to justify radiation treatment of the entire CNS in all patients? Second, in a particular patient is it possible to predict if the lesion will remain localized or disseminate? Data relative to these questions and to the control rate for different histologic grades of ependymomas are presented in Tables 3 and 4.

Table 3 shows the 5-year survival rates for several reported groups of patients. RINGERTZ and REYMOND (7) and CUSHING (8), who used surgical treatment only, reported 5-year survival rates of 27% and 20%, respectively. Survival rates for the UCSF series (9) and for the Columbia-Presbyterian Hospital, New York (10), including operative deaths, were 40% and 28%. If operative deaths and patients treated with irradiation doses of 3500 r or less are excluded, 5-year disease-free rates are 87%, 41%, and 58% respectively for the series of PHILLIPS, SHELINE, and BOLDREY (9), KRICHEFF et al. (10), and BOUCHARD and PEIRCE (11). Only 4 of the patients of KRICHEFF and co-authors and 1 of BOUCHARD and PEIRCE were irradiated to the full cerebrospinal axis. Interestingly, the only instance of spinal cord implant (10) was in a patient who received irradiation to the spinal cord. Firm conclusions cannot be drawn from these data; they suggest, however, that surgical treatment with adequate postoperative radiation therapy is the treatment of choice.

Table 3. Ependymoma survival rates (all ages)

Authors	Patients	Treatment	5-Year survival rate Supra-tentorial	Infra-tentorial	Total
RINGERTZ and REYMOND (7)	54	surgery only	15%	35%	27%
CUSHING[a]	25	surgery	-	-	20%
PHILLIPS, SHELINE, and BOLDREY (9)	35	only	31%	47%	40%
	9	≤ 3500 r	0%	0%	
	15	≥ 4500 r	80%	90%	87%
KRICHEFF et al. (10)	70		14%	33%	28%
	45	≥ 3600 r	22%	45%	41%
BOUCHARD and PEIRCE (11)	12	≥ 5000 r	-	-	58%

[a]As cited by DAVIDOFF (8).

Table 4. Survival rates for irradiated childhood ependymomas

Series	Number of patients	Unspecified 2 yr	Unspecified 3 yr	Histologic grade Low 5 yr	Histologic grade Low 10 yr	High 5 yr
ABRAMSON et al. (13) whole brain + some cords	20	30%	-	-	-	-
KRICHEFF et al. (10) primary site	36	-	25%	-	-	-
BLOOM (12) entire CNS	31	-	-	83%	80%	14%
SHELINE (3) primary site	14	-	-	78% (7/9)	72% (5/7)	20% (1/5)

Limited data available for childhood ependymomas treated by irradiation are summarized in Table 4. Lesions described as undifferentiated, ependymoma grade III or IV, and ependymoblastoma are classified in this table as high grade. Patients reported by BLOOM (12) received irradiation to the entire cerebrospinal axis. Those of ABRAMSON et al. (13) had radiation therapy to the whole brain and, in some instances, to the spinal cord as well. The patients of KRICHEFF et al. (10) and SHELINE (3) received irradiation only to the primary tumor site. The number of patients, particularly with data available regarding histologic grade, is small but the results indicate a high control rate for lesions of low histologic grade and a low survival rate for those of high grade irrespective of whether the spinal cord was irradiated. If any conclusion can be drawn, it is that local irradiation is sufficient and will control about 80% of low grade lesions. Because of the significant risk of spinal seeding in the high grade lesions, especially if they are infratentorial, it seems reasonable to irradiate the entire cerebrospinal axis in this group. Perhaps with current techniques of radiotherapy the salvage rate for the high grade ependymoma can be improved.

A recent review of the late results in children treated at UCSF (3) shows the quality of life in the survivors to be excellent. No neurologic deficits were produced by the radiation therapy; any such defects in the survivors were solely the result of initial damage from the tumor. Of the 5 patients who were treated in childhood and are now living 11 to 30 years post therapy, 3 are virtually normal. Another has a moderate hearing loss and tinnitus. The fifth is blind, which was his pretherapy condition.

OLIGODENDROGLIOMA

The reported UCSF (14) experience with oligodendrogliomas is limited to a total of 37 patients, all of whom underwent surgical procedures. Five of the 37 had operative deaths; sixteen received postoperative radiation therapy; sixteen had surgical treatment only. Review of the records disclosed no reason why some lesions were irradiated and others not. No correlation existed between the use of radiation and tumor location, histology, age of patient, duration of symptoms, postoperative condition of the patient, or year in which the patient was treated. Thus, for analysis of results, the group of patients who did receive radiation therapy appears to be comparable to that group which did not.

Table 5 shows the 5-year and 10-year survival rates. At 5 years the survival rate was 31% for those who received surgical treatment only and 85% for patients having surgical treatment plus irradiation. Application of the chi-square test indicated the difference to be statistically significant at the 5% level ($p = 0.02$). At 10 years, the survival rate for lesions irradiated was approximately double that for those nonirradiated; however, the number of patients available for 10-year follow-up study was small and the difference not statistically significant ($p = 0.4$). The median survival for the surgical patients was 2.8 years and for those treated by operation and irradiation 8.0 years. A recent follow-up study of the 6 irradiated patients, who at the time of our earlier reports were living at 10 years, showed that 1 subsequently died of recurrence and the other 5 are without evidence of recurrence (at 10, 13, 14, 21, and 22 years). Most survivors have maintained a useful and comfortable life for as long as we have follow-up information or until near the time of their death.

One possible, but unproved, instance of radiation necrosis occurred in a patient treated in the 1930's. The initial postoperative tumor dose was 5600 rads given in 30 days. Four years later an additional 5100 rads was given in 37 days. After the second course of irradiation, which was given for presumed tumor recurrence, the patient underwent an additional operation; however, recurrent tumor was not identified. The patient deteriorated progressively and died two years later. Unfortunately necropsy was not permitted and the cause of death is, therefore, uncertain.

Table 5. Results of treatment of 32 patients having oligodendrogliomas[a]

	Surgery only	Surgery + Irradiation
Number of patients	16	16
5-year survival	31% (4/13)	85% (11/13)
10-year survival	25% (2/8)	55% (6/11)

[a]Excludes 5 operative deaths.

It is concluded that, in this series, the use of postoperative radiation therapy yielded a significant increase in survival time and the quality of survival was good.

BRAIN STEM TUMORS

The next group of tumors is characterized by location rather than histology. These are tumors in the region of the thalamus, hypothalamus and brain stem where resection is not feasible and even biopsy is rarely attempted. They occur predominantly in children and most are thought to be gliomas.

Between 1949 and 1971, 41 children with such tumors were treated at UCSF (3). Only 4 were biopsied initially. In the remainder, diagnosis was based on history, clinical findings, and various radiographic contrast studies. All had pneumoencephalograms or ventriculograms or both. Many children also had arteriograms and several underwent Pantopaque[1] ventriculography.

The patients were divided into two groups. Group I included those with lesions in the region of the thalamus, hypothalmus, or floor of the third ventricle. While it is possible that some of the lesions were suprasellar germinomas, it is unlikely because none displayed the clinical triad of diabetes insipidus, visual disturbance, and hypopituitarism characteristic of the suprasellar germinoma. Group II included those patients with intrinsic lesions in the brain stem.

Data relative to these two groups of children and their initial response to radiation therapy are shown in Table 6. The average duration of symptoms was longer and the need for a shunting procedure more frequent in Group I. Three of the 4 patients, who had biopsy proof of glioma, are currently living without evidence of recurrence, their follow-up periods varying from 2 to 7 years. The patient with glioblastoma multiforme died within 4 months.

[1]Iophendylate injection U. S. P., manufactured by General Electric Company, Milwaukee, Wisconsin.

Table 6. Tumors of the third ventricle and brain stem

	Group I	Group II	Total
Patients (M/F)	14 (9/5)	27 (16/11)	41 (25/16)
Average duration of symptoms	8 mo	4 mo	5 mo
Biopsied	2	2	4
	1 astrocytoma	1 glioblastoma	
	NED (2yr)[a]	DOD[b] (4 mo)	
	1 astrocytoma	1 astrocytoma	
	NED (4 yr)	NED (7 yr)	
Surgically explored	10	13	23
Shunted	8	6	14
Improved post therapy	13	19	32 (78%)

[a]No evidence of disease (years postoperative).
[b]Died of the disease.

As indicated in the lower line of Table 6 about three-fourths of the 41 patients showed distinct and often considerable improvement during or shortly after completion of the therapy. In 21 the improvement was quite marked and in 9 others there was nearly complete disappearance of all neurologic signs. The initial response to irradiation is emphasized because it seems to lend credence to the presumptive diagnosis.

Table 7. Results of radiotherapy of tumors of the third ventricle and brain stem

	Group I	Group II	Total
Living, no evidence of disease (NED)			
present time	10/14	9/27	19/41
5 years post therapy	5/8 (63%)	10/24 (41%)	15/32 (48%)
10 years post therapy	3/5 (60%)	5/15 (33%)	8/20 (40%)
Dead of disease	4/14	18/27	22/41
survival range	8-24 mo	2 mo-8 yr	
median survival	8 mo	6 mo	
Autopsied	2	10	
Complications of irradiation	None	1 (absent GH,[a] NED 8 yr)	

[a]GH = growth hormone

Presently, 10 of the 14 patients of Group I and 9 of the 27 of Group II are living and without evidence of recurrence. The disease-free survival rates at 5 and 10 years after treatment are shown in Table 7. Group I did somewhat better than Group II, but the numbers are small and the difference is of questionable significance. The overall 5-year and 10-year survival rates were 48% and 40%, respectively. The quality of the survivors' lives has been excellent.

Twenty-two children died, 73% of them within the first year. The median survival for those dying of the tumor was 8 months in Group I and 6 months in Group II. It thus seems reasonable to view with optimism the life expectancy of the 15 patients who have survived beyond 5 years.

Postmortem examinations showed 3 glioblastoma multiforme, 7 astrocytomas, and 2 ependymomas, in addition to the 3 astrocytomas and 1 glioblastoma known from pretreatment biopsy.

To date only 1 patient has had a complication thought to be related to the radiation therapy. Because of previous belief that the normal pituitary is resistant to the usual therapeutic doses of irradiation, this complication is of particular interest. It occurred in a 5-year-old boy at the time of irradiation for a brain stem tumor. Five years after treatment, he stopped growing. This was not discovered until 1973, 8 years after treatment, at which time his serum growth hormone level was undetectable either upon fasting or following stimulation. All other pituitary hormone levels were normal, and the only clinical abnormality was that he was distinctly smaller than his siblings.

SUMMARY

The results of radiation therapy of astrocytomas, malignant gliomas, ependymomas, oligodendrogliomas, and intrinsic tumors in the wall or floor of the third ventricle and in the brain stem have been reviewed. Despite the many problems in analyzing results of treatment, it probably can be concluded that all of these gliomas, except glioblastoma multiforme, can be favorably influenced by irradiation with doses tolerated by the CNS; in some patients the lesion is either cured or, at least, controlled for many years. With a single course of radiation therapy as described the morbidity from irradiation has been minimal. In general, the quality of life for the survivor has been excellent.

REFERENCES

1. The Brain Tumor Study Group (under the auspices of the National Cancer Institute of the National Institutes of Health) Unpublished material.
2. BOUCHARD, J.: Textbook of Radiotherapy. 2nd ed., p. 366. G. H. FLETCHER 6ed.). Philadelphia: Lea & Febiger 1973.
3. SHELINE, G. E.: Radiation therapy of central nervous system tumors of childhood. In preparation.
4. MARSA, G. W., PROBERT, J. C., RUBINSTEIN, L. J., BAGSHAW, M. A.: Radiation therapy in the treatment of childhood astrocytic gliomas. Cancer 32, 646 (1973).
5. TAVERAS, J. M., THOMPSON, H. G., Jr., POOL, J. L.: Should we treat glioblastoma multiforme? Am. J. Roentgenol. Radium Therapy Nucl. Med. 87, 473 (1962).
6. KRAMER, S.: Cancer of the central nervous system: Radiation therapy in the management of malignant gliomas. In: Seventh National Cancer Conference Proceedings. American Cancer Society, Inc. 1973.
7. RINGERTZ, N., REYMOND, A.: Ependymomas and choroid plexus papillomas. J. Neuropathol. Exper. Neurol. 8, 355 (1949).
8. DAVIDOFF, L. M.: Thirteen year follow-up study of a series of cases of verified tumors of the brain. Arch. Neurol. Psychiat. 44, 1246 (1940).

9. PHILLIPS, T. L., SHELINE, G. E., BOLDREY, E.: Therapeutic considerations in tumors affecting the central nervous system: ependymomas. Radiology 83, 98 (1964).
10. KRICHEFF, I. I., BECKER, M., SCHNECK, S., TAVERAS, J. M.: Intracranial ependymomas. A study of survival in 65 cases treated by surgery and irradiation. Am. J. Roentgenol. Radium Therapy Nucl. Med. 91, 167 (1964).
11. BOUCHARD, J., PEIRCE, C. B.: Radiation therapy in the management of neoplasms of the central nervous system, with a special note in regard to children: Twenty years' experience, 1939-1958. Am. J. Roentgenol. Radium Therapy Nucl. Med. 84, 610 (1960).
12. BLOOM, H. J. G.: Treatment and prognosis of intracranial tumors in children. Presentation to the 26th Annual Midwinter Radiological Conference, Febr. 2. Los Angeles, Calif.: Los Angeles Radiological Society 1974.
13. ABRAMSON, N., RABEN, M., CAVANAUGH, P. J.: Brain tumors in children -- analysis of 136 cases. Unpublished presentation to the Radiologic Society of North America, Chicago, Ill., Nov. 1973.
14. SHELINE, G. E., BOLDREY, E., KARLSBERG, P., PHILLIPS, T. L.: Therapeutic considerations in tumors affecting the central nervous system: oligodendrogliomas. Radiology 82, 84 (1964).

Some New Radiotherapeutic Approaches and Combined Protocol Trials in the Management of Malignant Gliomas*

C. H. CHANG

The outlook for patients with malignant glioma (astrocytoma grade 3 and 4) is almost uniformly fatal. The length of survival following the conventional neuro-surgical intervention and radiotherapy can be measured in terms of months, a greater than 90% mortality within 18 months.

Considerable effort has been made by investigators in recent years to search for new avenues of approach. Among the new radiotherapeutic approaches, five major categories may be elucidated:

A. Optimal dose-response range and unconventional fractionations:

1. Search for optimal dose-response range.
2. Search for more effective regimens of dose-fractionation.

B. Radiosensitization:

1. Hyperbaric oxygenation.
2. Halogenated thymidine analogue (BUdR).
3. Hypermetabolic state.
4. Chemical sensitization of hypoxic cells.
5. Hyperthermia.

C. High LET radiations:

1. Fast neutrons
2. Pions

D. Combined radiotherapy & immunotherapy.

E. Combined radiotherapy & chemotherapy.

A. OPTIMAL DOSE-RESPONSE RANGE AND UNCONVENTIONAL FRACTIONATIONS

1. Search for Optimal Dose-Response Range

Although radiotherapy has been used in the treatment of glioblastoma for more than 50 years, no dose response curve for this particular tumor has been established.

In the 1950's BODEN (1) and LINDGREN (2, 3) began to define the maximal tolerance dose of normal brain tissue and spinal cord independently.

---

* This investigation was supported by USPHS research grant, No. CA-07962 from the National Cancer Institute.

In 1958, LINDGREN (3) quoted a first case on record in which glioblastoma had been destroyed by irradiation (Case of PENNYBAKER and RUSSELL (4)) and reported two cases of glioblastoma of his own. Both were given very high tumor doses (one received 7,250 R/27 days, the other 8,400 R/67 days in a "split-course" fashion with an initial course of 5,200 R/22 days and one month later a second course of 3,000 R/13 days). (Fig. 1.)

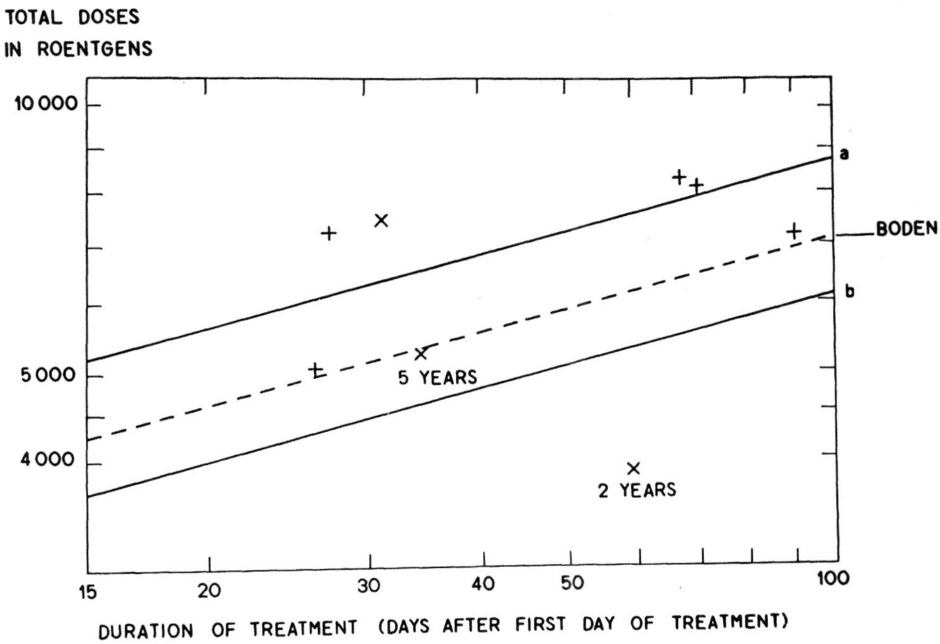

Fig. 1. Tumor doses in 5 cases of brain necrosis without residual tumors (+), in 3 cases with necrosis of normal tissue far outside of the tumor region (X) and their relation to tolerance dose levels. (The 2 glioblastoma cases by LINDGREN quoted in this paper designated as + are both located above the maximum tolerance line a). Courtesy of DR. M. LINDGREN and Acta Radiologica, (Suppl. 170, 1958)

It appears glioblastoma can be destroyed by a very high dose of radiation to an adequate treatment volume. The question is how high a risk of radiation necrosis one is willing to accept for the tumor sterilization. In order to deliver a very high dose with safety, the treatment volume must be reduced so that necrosis of adjacent normal tissues may be minimized.

CONCANNON and KRAMER (5) advocated whole brain irradiation based on their study of 50 cases of glioblastoma autopsy material. They found that a majority of the patients had an inadequate irradiation field and that in many cases, the tumor had invaded the brain stem.

On the other hand BLOOM et al.,(6) TODD et al., (7) and others advocate generous treatment volume instead of whole brain irradiation in order to achieve delivery of a desirably high dose and to reduce the calculated risk of complications.

For any clinical study, it would be of the first order of importance to determine the optimal radiation dose range to achieve tumor cell sterilization. At present it is quite acceptable to most radiotherapists that a tumor dose of 5-6000 rads be given in 5-7 weeks to the whole brain or to a generous tumor volume with a complication rate of about 5% for brain tissue necrosis. We may have to prepare to accept a complication rate of 10% or more if the tumor dose is to be raised to 7,000-7,500 rads in 8-9 weeks. Judging from the dismal survival figure for glioblastoma, it is justifiable to explore the upper end of a dose-response curve for glioblastoma in conventional radiotherapy using photons so that it may serve as a basis for future comparative studies using heavy particle therapy.

In a new glioblastoma protocol of R. T. O. G. (Radiation Therapy Oncology Group), one new study arm consisting of a higher tumor dose of 7,000 rads/8-9 weeks, (6,000 rads/7 weeks to whole brain + 1,000 rads/1-2 weeks as booster to cone-down tumor volume) is designed to compare with the conventional 6,000 rads/7 weeks (5,000 rads/5 weeks to whole brain and 1,000 rads/1-2 weeks as booster to tumor volume).

2. Search for More Effective Regimens of Dose-Fractionation

SIMPSON (8) and associates at the University of Toronto have conducted a randomized study using three radiation treatments per day, spacing every 8 hours including Saturday and Sunday for the treatment of malignant gliomas. The biological basis for this study is supposed to be that increase of the number of radiation exposures in intervals every 8 hours instead of every 24 hours may cope with that more rapidly proliferative fraction of malignant glioma cells. However, FOWLER (9) felt that shorter intervals than daily fractionation are only useful if specific peaks in the repair and progression time sequence can be hit reliably.

Three clinical groups were set up. The first group received a total dose of 4,000 rads given at treatment intervals of 8 hours for a period of 7 days. The second group had the same total dosage and same time interval but the treatments were spread over 21 days. The third group was given 4,000 rads with treatment every 24 hours spread over 21 days.

I. 4,000 rads - 7 days - 21 fractions at 8 hr. intervals.
II. 4,000 rads - 21 days - 63 fractions at 8 hr. intervals.
III. 4,000 rads - 21 days - 21 fractions at 24 hr. intervals.

They analyzed the results primarily in terms of the length of remission, which they considered the best index of the effectiveness of treatment. The curves for length of remission showed no significant difference between the three fractionation schedules (Fig. 2). Twenty two autopsy examinations have been carried out. In each of them tumor was present and none had recognizable radiation damage. They proposed to increase the dose to 4,250 rads. They felt that brain tissue can tolerate these higher dosage levels in the majority of patients.

B. RADIATION SENSITIZATION

1. Hyperbaric Oxygenation

Molecular oxygen has long been known as one of the most powerful dose modifying agents of the biological effect of ionizing radiations (10).

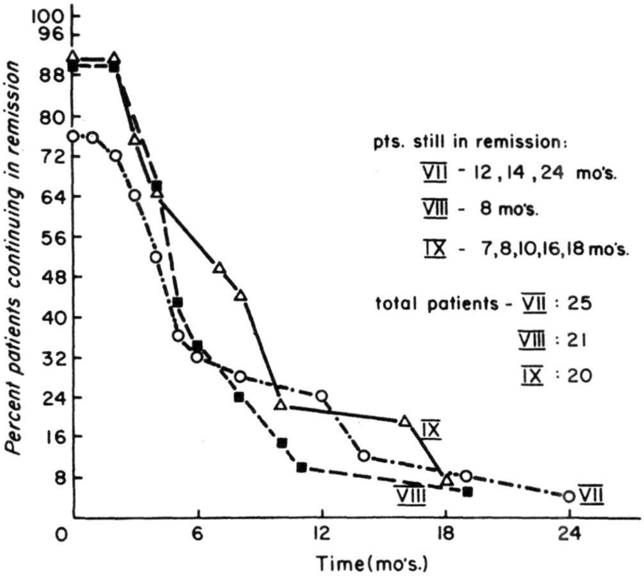

Fig. 2. Glioblastoma fractionation study: Length of remission of 3 study groups (excluding patients who did not complete radiation therapy). In the labels of the curves, curve VII represents group I, curve VIII group II, and curve IX group III, respectively as described in the text. Courtesy of DR. W. J. SIMPSON and BNL publication, (C-57), 1970

The survival curves for mammalian cells exposed to x-rays in the presence and absence of oxygen are illustrated (Fig. 3). The shape of the survival curve is the same under aerated and hypoxic conditions. All that is different is the magnitude of dose required to produce a given degree of biological damage. At all levels of survival the dose required under hypoxic conditions is about three times greater than the dose required under aerated conditions. This dose ratio between hypoxic condition and aerated condition is called oxygen enhancement ratio (OER).

One of the most practical ways to restore tissue oxygen tension is to breathe 100% oxygen in a pressurized chamber at 3 atmospheres (11). At this pressure, the solubility of O_2 in the plasma increases by about 20 fold.

CHANG (12) and associates at Columbia University - Presbyterian Medical Center started a pilot study in 1963 on radiotherapy of malignant gliomas under hyperbaric oxygenation with a pressurized transparent plastic chamber which contained pure oxygen at 3 atmospheres.

Glioblastoma multiforme was considered one of the unique malignant tumors for the clinical trial of the oxygen effect in radiotherapy because of (1) its uniformly rapid fatal course, (2) the absence of extracranial metastasis, (3) its abnormal tumor vasculature and (4) the high incidence of necrosis and hemorrhage in the tumor, whereby an anoxic or hypoxic tumor cell subpopulation most likely exists.

The initial dose schedule was 3,600 rads in 3 weeks i.e. 400 rads per treatment, three treatments per week (TIW), in a total of nine treat-

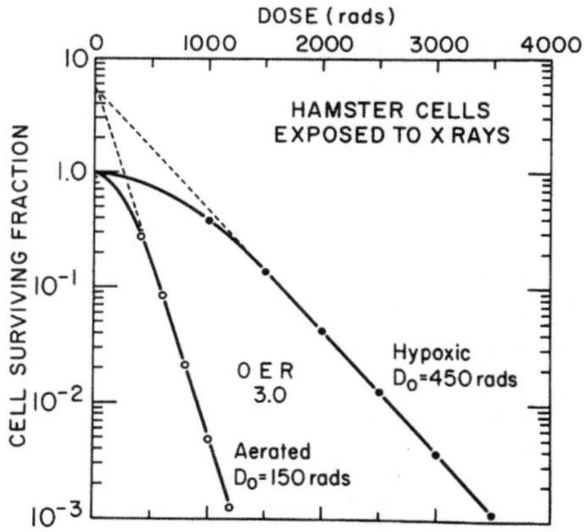

Fig. 3. Survival curves for cultured mammalian cells exposed to x-rays under aerated conditions, and under hypoxic conditions produced by passing a stream of purified nitrogen over the cells. These data are typical of many in the literature in which oxygen appears to be dose modifying. Courtesy of ERIC J. HALL, Ph. D. and Harper & Row, Publisher in Radiobiology for the Radiologist, PP. 52, 1973

ments, using radiocobalt source and bilateral opposing wide field technique. This was a conservative dosage to begin with but they wanted to avoid the potentially deleterious effect of combined oxygen and radiation in unconventional fractionation. The dose was systematically increased to 4,000 rads in 3 1/3 weeks, (400 rads, TIW); 4,400 rads in 3 2/3 weeks (400 rads TIW) and 5,000 rads in 5 weeks (333 rads TIW), as their experience with the oxygen toxicity and brain tissue tolerance increased. Since 1969, the dosage and fractionation have been changed to 6,000 rads in 6 weeks on a 5 treatments per week schedule. The initial 3,000 rads are delivered to the whole brain under conventional aerated condition and only the final 3,000 rads to a reduced volume encompassing the demonstrable tumor volume under hyperbaric oxygenation. This is because the biological knowledge of re-oxygenation taking place in tumors during daily fractionation has become well known. This pilot study was designed to include eventually a pair-arrangement technique whereby one in the pair was treated with pure oxygen breathing under 3 atmospheric pressures (30 PSI) and the other under conventional aerated condition as a control. This technique was followed as closely as possible but was not in strictly random fashion due to a difficult clinical situation in the early years of this study.

There were 82 patients admitted to the project, forty patients with hyperbaric oxygen and 42 as control. Two early patients in the oxygen group were excluded because the oxygen treatment was given for their recurrent glioblastomas several months after an initial course of radiotherapy (Table 1).

The results are analyzed according to survival rate, quality of survival and complications.

1. Survival Rate. An actuarial method of analysis is applied and the survival rate is depicted in two curves (Fig. 4). The curve A shows the survival rate of the oxygen study group (surgery plus radiotherapy plus oxygen) and curve B shows that of control group (surgery plus radiotherapy). The curve A shows a slightly higher survival rate at every

Table 1. Glioblastoma - Case distribution of various dose level in Oxygen Study Group and Air Control Group

Dose range / No. case / Study Group	Oxygen study group (HPO at 3 Atm.)	Air control group (Air at 1 Atm.)
3,600 rad/ 3 wk. (400 rad, TIW) or 1540 ret	8	10
4,000 rad/ 3 1/2 wk. (400 rad, TIW) or 1650 ret	4	4
4,400 rad/ 4 wk. (400 rad, TIW) or 1750 ret	2	5
5,000 rad/ 5 wk. (333 rad, TIW) or 1780 ret	6	8
6,000 rad/ 6-7 wk. (200 rad q. d.) 3,000 rad whole brain, 3,000 rad tumor volume, with or without oxygen. or 1770 ret	18	15
	38	42

3 month interval for the oxygen group than for the control group. The average 50% survival rate is about 2 months longer in the oxygen group than the control group. At the end of 18 months, the difference in survival rate appears most apparent, i.e. the survival rate for the oxygen group is about 28% and for the control group about 10%. At 24 months, the two curves begin to converse. However, we have 3 survivors in the oxygen group who are still alive at 25, 24 and 22 months respectively and 2 survivors in the control group still alive beyond 22 and 18 months respectively so that the shape of the later portions of these two curves cannot be predicted at this time. Because of small population samples, no statistical difference in survival rates can be drawn between the two groups.

2. Quality of Survival. Those who completed a course of radiotherapy under hyperbaric oxygen seemed to enjoy a better quality of survival than those in the radiotherapy control group. For example, one patient who was a manager of a supermarket, was able to return to his job in full capacity for at least 6 months following a course of radiotherapy under hyperbaric oxygen. Another patient who was a banker, returned to work for 1 year before he was advised to slow down.

3. The Complications. Most patients tolerated the hyperbaric oxygen procedure well. Some of them continued their treatments as out-patients after having received initial treatments for one or two weeks in the hospital. Only two patients developed generalized convulsions at the time of decompression due to oxygen toxicity. The incidence is less than 1/2% in terms of the total number of exposures in our series. Several patients developed mild otitis media following myringotomy and

hyperbaric therapy. One patient developed a severe skin reaction with subsequent disruption of the scalp wound. Three patients had extensive radiation necrosis of the brain observed at a subsequent "second look" surgery or at autopsy.

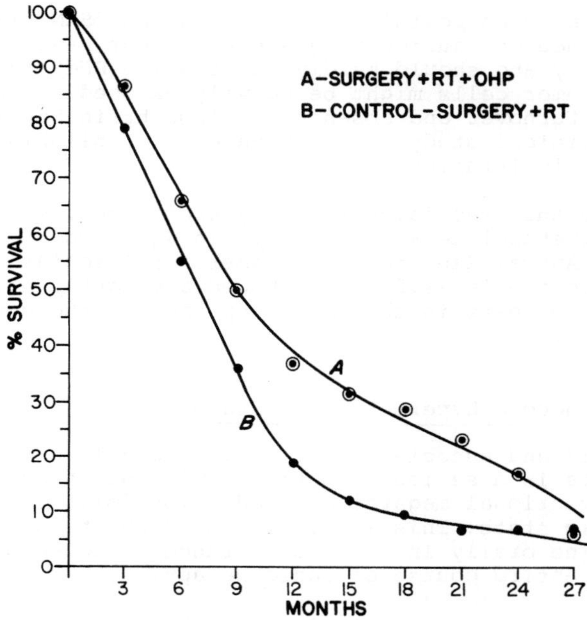

Fig. 4. Glioblastoma-post radiotherapy survival rate. Curve A represents the study group (surgery plus radiotherapy plus hyperbaric oxygen) and consists of 38 primarily treated cases. Curve B represents the control group (surgery plus radiotherapy only) and consists of 42 primarily treated cases. It appears that the difference in survival rate between the two groups is most apparent at 15-18 months time after therapy and that the survival for the control group at this time is only about 10% while that for oxygen study group is about 25-30%

This preliminary report represents a prospective clinical study of pilot nature. During the course of this study, we have observed microscopically that a glioblastoma could be completely destroyed by irradiation under hyperbaric oxygen, probably in the dose level of 4,000 rads in 3 1/3 weeks and above in 3 treatments per week schedule (NSD $\geq$ 1650 ret). The autopsy material of one patient with glioblastoma showed a complete disappearance of tumor cells in the irradiated post-operative tumor field. However, the radionecrosis of the adjacent normal brain tissue appeared to be very extensive.

It is felt that this type of clinical study on oxygen effect is paving the way for future study with high L. E. T. (Linear Energy Transfer) radiation. One of the advantages of high L. E. T. radiation over low L. E. T. radiation (Photons) is its lower oxygen enhancement ratio.

2. Halogenated Thymidine Analogues

DJORDJEVIC and SZYBALSKI (13) in 1960 and KAPLAN (14) in 1962 showed that 5-Bromo-deoxyuridine (BUdR), an thymidine analogue, incorporating in DNA in dividing cells in tissue culture increases the sensitivity of these cells by two to three times on single exposure to radiation. These incorporated halogenated pyrimidines interfere with the formation of purine-pyrimidine base linkages thereby changing the forces that bind the two DNA strands together (15).

It has been postulated by HOSHINO and SANO (16) that normal glial cells and neurons surrounding a brain tumor have essentially no mitotic activity and should hardly incorporate BUdR into their nuclear DNA, whereas tumor cells might be heavily exposed to the BUdR due to rapid tumor DNA turnover and absence of blood brain barrier. They have embarked on a clinical study in the treatment of glioblastoma with continuous intracarotid infusion of BUdR.

SANO and associates recently presented the results of BAR therapy (BUdR-Antimetabolite-Radiotherapy) in 42 cases of highly malignant gliomas at the Annual Meeting of the American Association of Neurological Surgeons in Boston in 1972. They showed a slightly higher survival at the end of two years in the BAR group than in the conventional radiotherapy group.

3. Induced Hypermetabolic State

GRIEM and associates (17) have recently presented the preliminary results in a series of patients with glioblastoma who were treated with conventional megavoltage radiation following induction of a hypermetabolic state. This is accomplished by the administration of tri-iodothyronine orally in the dosage range of 300-600 microgram per day during the entire course of radiotherapy.

A few patients survived beyond 5 years with such combined treatment. This hyper-metabolic approach may have sensitized tumor cells and deserves further investigation.

4. Chemical Sensitization of Hypoxic Cells

Chemicals which selectively radiosensitize hypoxic cells have been considered by ADAMS (18) as adjuvants in the radiotherapy of human tumors. CHAPMAN and associates (19) have screened various nitroheterocyclic drugs and have indicated that some nitrofuran derivatives are active as sensitizers in mammalian cells *in vitro*. The radiosensitizing effect is primarily a radiation dose modifying effect. Metronidazole, 2-nitro-imidazole, etc. have been found to be effective in enhancement of radiation response in solid tumors at the Gray laboratory (20). The enhancement ratio of 1.3 was obtained for metronidazole (Flagyl) and the ratio of 1.6-2.1 were found using a 2-nitro-imidazole drug. A clinical trial has been begun at Mount Vernon Hospital, England. A successful application of such drugs in the radiotherapy of human tumors including glioblastoma would depend upon pharmacological properties such as toxicity of the drug, concentration of the drug in blood and tissue fluid and the accessibility of hypoxic cells to the drug.

5. Hyperthermia

Interest in application of hyperthermia in potentiation of radiation response has recently been revived. BEN-HUR *et al.* (21) have provided sound laboratory evidence in supporting such endeavor (Fig.5). It appears that such enhancement is due to inhibition of repair of sublethal

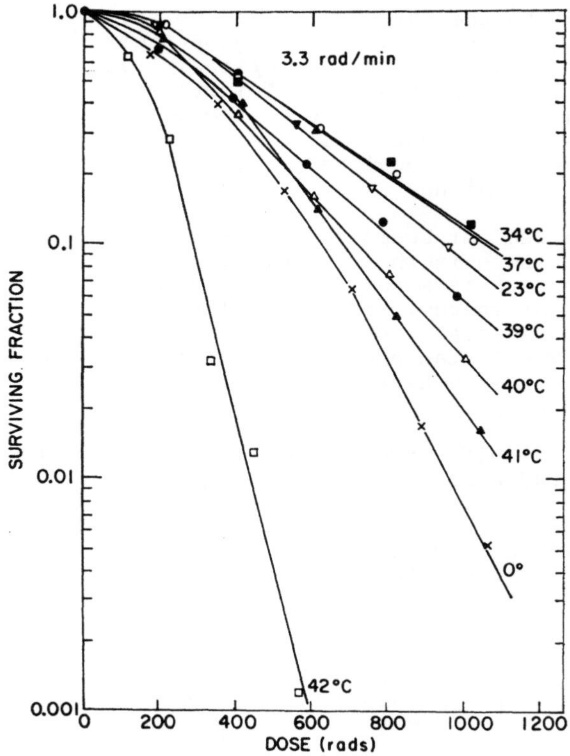

Fig. 5. Survival of Chinese hamster cells x-irradiated while suspended in DUM-20 at different temperatures (250 kvp, 3.3 rad/min.). 0°C, x; 23°C, ▽; 34°C, ■; 37°C, O; 39°C, ●; 40°C, △; 41°C, ▲. Lines are fitted by eye. Courtesy of DRS. BEN-HUR, ELKIND and BRONK and Academic Press. (Radiation Res. 58, 38, 1974)

damage and enhancement of lethal damage. It is felt that the brain tumor temperature could be raised during the time of radiation exposure by an ultrasound or microwave technique. Many technical difficulties should be overcome before it can be applied to the clinical situation, such as cerebral glioblastoma.

C. HIGH LET (LINEAR ENERGY TRANSFER) RADIATIONS

In recent years considerable attention has been given to the study of particulate radiation (high LET range) in the treatment of malignancy. Application of heavy particles, such as neutrons and pions have been suggested or tried in the management of intracranial neoplasms because of the special characteristics of this type of radiation.

In order to understand the rationale underlying their utility, it is desirable to summarize the physical and radiobiological characteristics of heavy particle beams.

High energy heavy particle beams can be sharply collimated. They show little side scattering and are highly penetrating with little energy lost to the surrounding medium early in their path. They have a definite range in tissue, depending on their initial energy. As the particles slow down near the end of their path, they release even greater amounts of energy in the form of increased ionization. The increased density of ionization at the end of the path, known as the Bragg peak, has particular significance for radiobiological reasons as well as for making it possible to localize precisely the region of high dose while sparing other regions.

The radiant beams exert their biological effects by release of energy to tissue in the form of ionization. The density of ionization along the path of a particle or photon is related to this release or transfer of energy, and is usually expressed as linear energy transfer (LET). LET is usually expressed in units of kilo-electron volts per micron, (KEV/u). Standard 200 kvp X-rays have an LET value of about 2 kev/u. High energy X-rays are less densely ionizing and have an LET of about 0.5 kev/u.

The significance of the LET is its relationship to the biological effects of radiation (22). The RBE (relative biological effect) generally increases with increasing LET up to about 110 kev/u. Beyond this point, there is no further biological effect because of saturation or "overkill" of high LET (Fig. 6). The relation of the LET to the oxygen factor in radiotherapy can be stated that as LET increases, the oxygen enhancement ratio (OER) reduces.

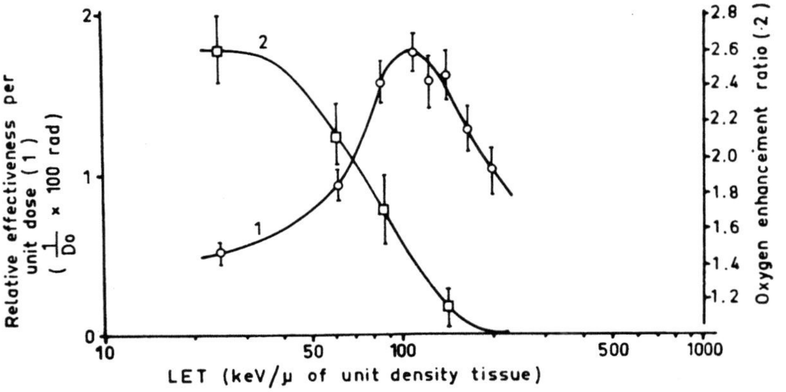

Fig. 6. Relation of LET to relative biological effect and to the oxygen factor in radiosensitivity. Curve 1, variation of RBE with LET. Curve 2, variation of oxygen enhancement ratio with LET. Courtesy of G. W. BARENDSEN, Ph. D. and the Williams & Wilkins Co. Baltimore. In Cellular Radiation Biology. PP. 333, 1965

In the cell survival curve of low LET radiation, there is a shoulder in the initial low dose region, which indicates a cumulative effect of perhaps several "hits" being required to affect each target area or indicates a repair process taking place in sublethal damage. If one irradiates a cell culture to a given dose, waits 24 hours, and then repeats the irradiation, the shoulder is again seen following the second irradiation dose, indicating complete recovery from any sublethal irradiation effect. On the other hand, heavy particle beams with their high ionization density show almost no cumulative shoulder effect. The curve is nearly truly exponential, no recovery is evident and repeated fractions of dose are additive (Fig. 7). The RBE appears to be much higher when the treatment is fractionated into multiple small doses.

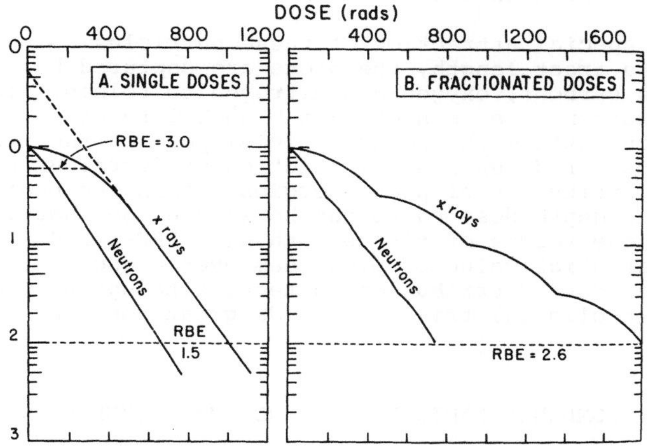

Fig. 7. Panel A: single dose. Typical survival curves for mammalian cells exposed to x-rays and fast neutrons. In the case of x-rays the survival curve has a large initial shoulder; for fast neutrons the initial shoulder is smaller and the final slope steeper. The two survival curves have different shapes. Panel B: fractionated doses. Illustration of the effect of giving doses of x-rays or fast neutrons in four equal fractions to produce the same level of survival as in panel A. The shoulder of the survival curve is reexpressed after each dose fraction; since the shoulder is larger for x-rays than for neutrons, this results in an enlarged RBE for fractionated treatments. Courtesy of ERIC J. HALL, Ph. D. and Harper & Row, Publisher. In Radiobiology for the Radiologists, PP. 68, 1973

The discovery of neutrons by CHADWICK in 1932 and early experiments led to a clinical trial under the direction of DR. ROBERT STONE in the late 1930's at the University of California. Early effects on tumors which were irradiated appeared encouraging, but marked late fibrosis and necrosis in normal tissues caused a cessation of the trial.

With increasing radiobiological knowledge, the question of fast neutron therapy was reopened in the 1960's. A series of pretherapeutic biological experiments has been performed and clinical trials are being under way in Europe and in this country.

1. Fast Neutron. A 6 Mev. neutron beam (produced by a British Medical Research Council's Cyclotron at Hammersmith Hospital in London) has a depth dose distribution intermediate between that for 250 K. V. X-ray and ^{60}Co. At 30 Mev (produced by a cyclotron at Texas A & M) a neutron beam has a depth dose curve which is slightly superior to that of a ^{60}Co beam.

Clinical trials with neutron beams for the treatment of glioblastoma have been started in Hammersmith (by DRS. CATTERALL and BLOOM), in Texas (by DR. FLETHER), in Washington University, Seattle (by DR. PARKER) and in the Naval Research Laboratory, Virginia (by DR. ROGER). Each group has treated about 5 to 15 cases up-to-date and it is too early to evaluate the results.

2. Pion-Negative Pi-Meson. This particle, heavier than an electron but lighter than a proton (mass, 276 times the electron) can be produced in the synchrocyclotron by interaction of protons with a beryllium target and then separated electromagnetically.

A special attraction of these particles is the fact that at the end of the tract length, the pions are captured by light atomic nuclei, such as carbons, oxygen and nitrogen to release alpha particle, proton and neutron, i.e. a number of high LET reaction products (Bragg peak). This contrasts with the superficial part of the tract length where energy loss produces low LET reaction products. By careful planning of the distribution of pion energies within the beam, the high LET region of the depth dose curve can be widened so that it is feasible to put a large volume of tissue with high RBE-low OER in part of the beam. The potential value of pion beam over a good neutron beam lies in the better dose distribution pattern. Although no pion beam is yet available for clinical trial, it holds great hope for the near future.

D. COMBINED RADIOTHERAPY AND IMMUNOTHERAPY

It has been generally conceived that surgery and radiotherapy will take care of large tumors per se and chemotherapy or immunotherapy will take care of tumor residua or micro-metastases.

BLOOM and associates (6) have conducted a combined radiotherapy and immunotherapy randomized clinical trial to assess the value of specific active immunotherapy using heavily irradiated autologous tumor cells in patients with glioblastoma treated by radical surgery and post-operative irradiation.

The results of 62 patients so treated showed no statistically significant difference in survival between the group receiving adjuvant autologous tumor cells and those treated with surgery and radiation alone (Fig. 8). The results were considered sufficiently discouraging to abandon this clinical trial at this stage.

Other clinical trials for glioblastoma utilizing combined radiotherapy immunotherapy principles have not yet appeared in the literature.

E. RADIOTHERAPY AND CHEMOTHERAPY - A COMBINED PROSPECTIVE PROTOCOL TRIAL

Chemotherapy of malignant brain tumors thus far has been disappointing (26). In previous uncontrolled studies chemotherapy with a variety of agents failed to provide evidence of improvement in the treatment of glioblastoma. Mithramycin (25 mcg/kg./day x 21 days, IV) is the only drug to be employed in a prospective, controlled, randomized study by the Brain Tumor Study Group (BTSG) and was shown to be of no distinct value in the treatment of gliomas when compared with untreated patients (27).

With the advent of a new class of chemotherapeutic agents, nitrosourea which is lipid soluble to allow it to cross the blood brain barrier and which has demonstrated efficacy in many animal and human tumors, the Brain Tumor Study Group has conducted a prospective randomized study

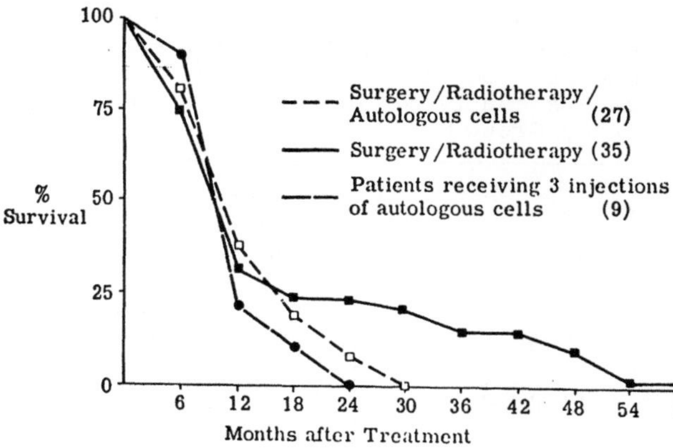

Fig. 8. Glioblastoma: Survival of trial patients according to treatment category. Survival of patients receiving 3 injections of autologous cells is shown separately. Courtesy of DRS. H. J. G. BLOOM et al. and Brit. J. Cancer 27, 253 (1973)

since 1969. The preliminary analysis (28) tends to indicate a significant longer median survival time in patients with glioblastoma receiving radiotherapy post-operatively than in patients receiving no radiotherapy, and that BCNU plus irradiation may be of greater value than either alone.

In a proposed prospective, randomized protocol for malignant astrocytoma, the Eastern Cooperative Oncology Group (ECOG) has selected DTIC (Imidazole Carboxamide) as one of the study arms based on their previous encouraging experience with this drug in a phase II screening study (29) and their treatment of 15 patients with DTIC and whole brain irradiation. The median survival (30) for the 15 patients is 55 weeks with 6 patients still alive.

In order to develop a meaningful combined protocol for the Radiation Therapy Oncology Group (RTOG) which is primarily concerned with radiation parameters, it is felt it should include one study arm similar to that of the Brain Tumor Study Group (Radiation plus BCNU) and another arm similar to that of the Eastern Cooperative Oncology Group (Radiation plus DTIC plus MeCCNU), so that intergroup comparison of results among these three major Oncology groups will be possible. A combined prospective randomized protocol for malignant Gliomas has been approved by both RTOG and ECOG groups and a schematic diagram of the combined protocol appears as follows: (Table 2)

Accession of cases by these two groups is estimated to be 150 per year and it is expected that in 3 years, enough cases will be accrued for a meaningful evaluation. Meanwhile, the results will be compared with those of the Brain Tumor Study Group in which one study option (BCNU) is in common.

SUMMARY

This paper presents a comprehensive review of new radiotherapeutic approaches with radiation alone or in combination with other therapeutic disciplines, in the treatment of glioblastomas.

Table 2. Glioblastoma protocol - option schema (RTOG - ECOG). This joint protocol between RTOG (Radiation Therapy Oncology Group) and ECOG (Eastern Cooperative Oncology Group) is so designed that the treatment results can be valid for intergroup comparison with BTSG (Brain Tumor Study Group)

Stratification by 1. Institution Surgery ....→ 2. Histology (to be recorded only) 3. Extent of Surgery (to be recorded only)	R A N D O M I Z A T I O N	1	IRRADIATION 5000 rads/5 weeks Whole Brain 1000 rads/1-2 weeks Booster - One Course
		2	IRRADIATION 6000 rads/7 weeks Whole Brain 1000 rads/1-2 weeks Booster - One Course
		3	IRRADIATION 6000 rads/7 weeks Whole Brain- One Course BCNU 80 Mg./M^2/Day x 3 -Q.6-8 week
		4	IRRADIATION 6000 rads/7 weeks Whole Brain- One Course MeCCNU 150 Mg./M^2/Day x 1 -Q.8 week DTIC 175 Mg./M^2/Day x 5 -Q.4 weeks
			*PILOT STUDY - BUdR Irradiation-6000 rads/7 weeks - One Course BUdR -1.0 Gm. Q. D. Intra-carotid infusion - One Course

Recent advances in radiotherapy can be summarized along 3 main lines:

1. Increasing knowledge and application of radiobiological principles for more effective regimens of dose-fractionation and for enhancement of radiation sensitivity, such as oxygen effect and repair of sublethal radiation damage.
2. Development and availability of high LET radiation sources for clinical radiotherapy.
3. Combined radiotherapy and chemotherapy or immunotherapy.

An exploration of a new dose-fractionation regimen (3 treatments per day) for glioblastoma has been carried out by SIMPSON and associates at the University of Toronto. Determination of the upper portion of the dose-response curve of glioblastoma (7,000 rads/8-9 weeks) for obtaining more tumor cell sterilization has been designed for clinical trial by RTOG group.

CHANG and associates have conducted a clinical trial using hyperbaric oxygen for sensitization of hypoxic glial tumor cells. A slightly higher survival rate at 18 months level and better quality of survival have been noted. However, due to the small number of cases, the difference is not statistically significant.

SANO and associates have conducted a clinical trial using BUdR, in intracarotid infusion in order to achieve radiosensitization via interference with formation of pyrine-pyrimidine base linkage.

GRIEM and associates have induced a hypermetabolic state of the patient with glioblastoma by administration of tri-iodothyronine during the entire course of radiotherapy. ADAMS and others have attempted to use chemicals to sensitize the hypoxic tumor cells selectively. The promise of high LET radiation lies in its biological characteristics of low oxygen enhancement ratio and absence of repair of sublethal damage. Clinical trials for glioblastoma using neutrons have been initiated in Europe and in the U. S. A.

The results of combined radiotherapy and immunotherapy for glioblastoma as reported by BLOOM and associates have been disappointing.

The preliminary results of combined radiotherapy and chemotherapy with nitrosourea and DTIC have been encouraging. Two combined radiotherapy and chemotherapy protocols for clinical trials are being conducted by BTSG and RTOG-ECOG respectively. The protocols are so designed that their results can be valid for intergroup comparison.

Although the results of treatment of glioblastomas have been dismal in the past, the outlook appears to be more encouraging with these new radiotherapeutic and combined chemotherapeutic approaches.

REFERENCES

1. BODEN, G.: Radiation Myelitis of brain-stem. J. Radiologists 2, 79 (1950).
2. BERG, N. O., LINDGREN, M.: Time-Dose relationship and morphology of delayed radiation lesions of the brain in rabbits. Acta Radiol. Suppl. 167 (1958).
3. LINDGREN, M.: On tolerance of Brain Tissue and Sensitivity of Brain Tumors to Irradiation. Acta Radiol. Suppl. 170 (1958).
4. PENNYBAKER, J., RUSSELL, D. S.: Necrosis of the Brain due to Radiation Therapy. J. Neurol. Neurosurg. Psychiat. 11, 183 (1948).
5. CONCANNON, J. P., KRAMER, S., BERRY, R.: The Extent of Intracranial Gliomas at Autopsy and its Relationship to Techniques Used in Radiation Therapy of Brain Tumors. Am. J. Roentgenol. 84, 99 (1960).
6. BLOOM, H. J. G., PECKHAM, M. J., RICHARDSON, A. E., ALEXANDER, P. A., PAYNE, P. M.: Glioblastoma Multiforme: A Controlled Trial to Assess the Value of Specific Active Immunotherapy in Patient Treated by Radical Surgery and Radiotherapy. Brit. J. Cancer 27, 253 (1973).
7. TODD, I. D. H.: Choice of Volume in the X-ray Treatment of Supratentorial Gliomas. Brit. J. Radiol. 36, 645-649 (1963).
8. SIMPSON, W. J.: In Time and Dose Relationships in Radiation Biology as Applied to Radiotherapy. BNL 59203 (C-57) P. 301-302. Carmel, Calif.: NCI-AEC Conference 1970.
9. FOWLER, J. F.: The Rational of Dose Fractionation. Front. Radiation Ther. Onc. 3, 6 (1968).
10. GRAY, L. H., CONGER, A.D., EBERT, M., HORNSEY, S., SCOTT, O. C. A. "The concentration of oxygen dissolved in tissues at the time of irradiation as a factor in radiotherapy". Brit. J. Radiol. 26, 638 (1953).
11. CHURCHILL-DAVIDSON, I., SANGER, C., THOMLINSON, R. "Oxygenation in radiotherapy. II: Clinical application". Brit. J. Radiol. 30, 406 (1957).
12. CHANG, C. H., SEAMAN, W. B., JACOX, H. W.: Clinical Aspects of Hyperbaric Oxygen and Radiotherapy - New York Experience. Front. Radiation Ther. Onc., 1, 183 (1968) and (unpublished data).
13. DJORDJEVIC, B., SZYBALSKI, W.: Genetics of human cell lines. III.

Incorporation of 5-bromo- and 5 deoxyuridine into the deoxyribonucleic acid of human cells and its effect on radiation sensitivity. J. exp. Med. 112, 509 (1960).
14. KAPLAN, H. S., SMITH, K. C., TOMLIN, P.: Effect of halogenated pyrimidines on radiosensitivity of E. coli. Radiation Res. 16, 98 (1962).
15. WAKE, R. G., BALDWIN, R. L.: Physical studies on the replication of DNA in vitro. J. mol. Biol. 5, 201 (1962).
16. HOSHINO, T., SANO, K.: Radiosensitization of Malignant Brain Tumors with Bromouridine (Thymidine Analogue). Acta Radiol. 8, 15 (1969).
17. GRIEM, M., Associates: A paper presented at the 21st Annual Meeting of Association of University Radiologists, Vancouver, Canada, May 1973 and personal communication 1974.
18. ADAMS, G. E.: Chemical Radiosensitization of Hypoxic Cells. Brit. Med. Bulletin 29, 48 (1973).
19. CHAPMAN, J. D., REUVERS, A. P., BORSA, J., PETKAU, A., MCCALLA, D. R.: Nitrofurans as Radiosensitizers of Hypoxic Mammalian Cells. Cancer Res. 32, 2616 (1972).
20. Annual Report of Gray Laboratory, Cancer Research Campaign, Mount Vernon Hospital, Northwood, England. Director, J. F. FOWLER, Ph. D., 1973.
21. BEN-HUR, E., ELKIND, M. M., BRONK, B. V.: Thermally enhanced Radioresponse of Cultured Chinese Hamster Cells: Inhibition of Repair of Sublethal Damage and Enhancement of Lethal Damage. Radiation Res. 58, 38 (1974).
22. BARENDSEN, G. W.: The Influence of Oxygen on Damage to the Proliferative Capacity of Cultured Human Cells Produced by Radiations of Different LET in Cellular Radiobiology. Edited by University of Texas M. D. Anderson Hospital and Tumor Institute. Baltimore: William and Wilkins 1965.
23. BOND, V. P.: Negative Pions: Their Possible Use in Radiotherapy. Am. J. Roentgenol. 111, 9 (1971).
24. SHAPIRO, W. R., AUSMAN, J. I.: The Chemotherapy of Brain Tumors: a Clinical and Experimental Review. In Recent Advances in Neurology. Edited by F. PLUM. Philadelphia: F. A. Davis Co. 1969.
25. LEVENTHAL, C. M., WALKER, M. D.: Chemotherapy of Malignant Glioma: a Collaborative Study. Excerpta Medica Int. Congress Series. 193, 33 (1969).
26. WALKER, M. D., GEHAN, E. A.: An Evaluation of BCNU and Irradiation Alone and in Combination for the Treatment of Malignant Glioma. Proc. Am. Ass. Cancer Res. 13, 67 (1962).
27. KINGRA, G., COMIS, R., OLSON, K., HORTON, J.: Imidazole-4 (or 5) Carboxamide, 5 (or 4)-(3,3 dimethyl-1-trizeno), (DTIC) NSC-45388 in the Treatment of Malignant Tumors other than Melanoma. Cancer Chemotherapy Rep. 55, 281 (1971).
28. HORTON, J., CUNNINGHAM, T.: Unpublished data at Albany Medical College and personal communication 1974.

Discussion

M. L. GRIEM and D. H. PEARSON

This portion of the symposium deals with the treatment of cerebral gliomas by the nonsurgical approach to therapy. Dr. BUCY has pointed out that surgical removal of the more malignant gliomas is not feasible. Total extirpation of the tumor leaves the patient with an unacceptable outcome in most treatment situations.

There are two alternate possibilities, irradiation and chemotherapy. There may be combinations of irradiation and chemotherapy which would yield optimum results. MENDELSOHN has pointed out the problems facing the radiation therapist and the chemotherapist (4). There is a family of dose-response curves for the ablation of tumor and for the complications associated with normal tissue response. These curves indicate the ability to ablate the tumor. Similarly the response of normal tissue to drugs and radiation can be recorded as a function of increasing dose of agent. With this data, MENDELSOHN has pointed out the use of game theory to help in medical decision making. He has generated Therapeutic Operating Characteristic (TOC) curves and from these TOC curves a therapist can select the optimum to yield the best possible result with the fewest complications. If one considers the outcome of such decision making and its effect on the patient and on society, one can hope to see improved results of treatment.

The radiation therapist, like the surgeon, is able to produce local control of malignant disease. The chemotherapist, on the other hand, is able to control disease in multiple sites by systemic administration of chemotherapeutic agents. There is interaction between one injurious agent and another. By knowing more of the biology of the treatment situation, the therapist might be able to maximize the injury in the tumor without seeing concomitant and equal injury in normal tissue. In order to improve the treatment situation much more knowledge is needed concerning the biology of the tumor and the surrounding brain and connective tissue. We need to know more about cell kinetics and the growth fraction of the tumor. Little is known about the level of tissue oxygenation in the tumor, a factor important in radiation therapy and probably an important factor in chemotherapy. We know little about the normal cell turnover of the brain and the connective tissue and blood vessels of the brain.

Certain aspects of the information presented by Dr. KIRSCH concern me. Early in his presentation he has used the colchicine accumulation method to study cell kinetics. VAN'T HOF (6) has demonstrated in tissue culture that colchicine may not produce an absolute block and that data so derived may have certain errors which may be related to the dose of drug which is available within the cell. An excellent review of the radiation sensitivity and drug sensitivity of cells has been made by MADOC-JONES (3). Several investigators including STEWARD, HAHN (5), and

COHEN (1), have utilized such data generated by MADOC-JONES and others to predict the optimum utilization of drugs and radiation. This is possible provided one knows the details of the biology of tumor and normal tissue growth.

Recently A. YUNG at the University of Chicago has analyzed the treatment of glioblastoma multiforme patients using radiation therapy and an induced hyperthyroid state (8). Analysis of these patients indicated improved survival over those patients treated with radiation therapy alone. It is interesting to speculate why the induced hyperthyroid state might be operating in a beneficial effect. The cardiac output is increased and blood flow throughout the body is enhanced. Possibly the degree of blood flow in the tumor is increased providing increased oxygenation of the tumor. GOODMAN and GILLMAN (2) suggest that the normal brain is unaffected by thyroid drugs and the effects of thyroid are principally on the oxidative phosphorolation within the cell. It may also be that the induced hyperthyroid condition recruits from a resting state into a proliferating state thereby making those cells more sensitive to ionizing radiation.

WITHERS (7) has outlined four "R's" when considering tumors and normal tissue in the treatment situation. These are sublethal repair, redistribution, repopulation, and reoxygenation. Recently he has added a fifth "R" which he has termed rejection. The role of the immune system in tumor rejection may also be a very important factor in the treatment of cerebral gliomas, yet we know little about the immune system within the brain.

REFERENCES

1. COHEN, LIONEL: Cell population kinetics in radiation therapy: optimization of tumor dosage. Cancer 32, 236 (1973).
2. GOODMAN, L. X., GILMAN, A.: The Pharmacological Basis of Therapeutics: A Textbook of Pharmacology, Toxicology, and Therapeutics for Physicians and Medical Students. Second edition, chapter 65, pp. 1528-1565. New York: The Macmillan Company 1955.
3. MADOC-JONES, H., MAURO, F.: Site of Action of Cytotoxic Agents in the Cell Life Cycle. Handbook of Experimental Pharmacology. New Series, chapter 10, pp. 205-219. Berlin, Heidelberg, New York: Springer 1974.
4. MOORE, D. H. II, MENDELSOHN, M. L.: Optimal treatment levels in cancer therapy. Cancer 30 (1972).
5. STEWARD, P. G., HAHN, G. M.: The application of age response functions to the optimization of treatment schedules. Cell and Tissue Kinetics 4(1971).
6. VAN'T HOF, J., WILSON, G. B., COLON, A.: Studies on control of mitotic activity: use of colchicine in tagging of synchronous population of cells in meristem of Pisem sativum. Chromosoma 1960, II, 313-321.
7. WITHERS, H. R.: Capacity for repair in cells of normal and malignant tissues. Time and Dose Relationships in Radiation Biology as Applied to Radiotherapy. (NCI-AEC Conference, Carmel, California, 1969). BNL 50203 (C-57). Upton, New York: Brookhaven National Laboratory 1970.
8. YUNG, WAI-KWAN, STEWARD, WILLIAM, MARKES, JAMES E., GRIEM, MELVIN L., MULLAN, JOHN F.: Glioblastoma multiforme: treatment with radiation and triiodothyronine. Submitted to Journal of Neurosurgery.

Closing Remarks

H. M. Zimmerman

It would be impossible in the 15 minutes allocated for the summation of this elegant meeting to do justice in summarizing each of the papers. Nor is this necessary for the reason that these will be published, and I hope soon, and each of the speakers I am sure will summarize what he has said, so that I will confine myself to brief remarks in the way of addenda to what has been said. I might add that all of the speakers, including myself, who came to teach have stayed to learn. This I think was an excellent symposium and I cannot thank the organizers enough for it.

The introduction on experimental brain tumors, which I attempted so ineffectively to present yesterday, had to do with the attempt to reproduce tumors in the experimental animal that could be dealt with in large numbers instead of waiting for the occasional human glioma to appear. The methods employed for the production of these tumors utilized the aromatic hydrocarbons methylcholanthrene, benzpyrene and dibenzanthracene. The more recent studies with methyl and ethyl nitrosourea have the advantage over the pellet implantation method employed by the older workers in that one can inject these substances intravenously and even feed them to animals. And I remember DRUCKREY who started this nitrosourea method of chemical carcinogenesis saying that he was trying to reproduce something that would be more etiologically pertinent than the hydrocarbons. That is, probable dietary impurities of the nitrosourea variety could be ingested by man and possibly give rise to neoplasms. The disadvantage of course with the resorptive carcinogens is that you never can tell in advance where the tumors will arise, and I've been surprised once when autopsying a rat after several injections with nitrosourea in finding six different gliomas in different parts of the nervous system, two in the spinal cord and four in the brain, and several of them were of different kinds so that they were not metastatic lesions from one spot to another. The localization that is possible with the implantation of chemical carcinogens makes you more certain of what the lesion is. You are dealing with one lesion at a time that can be transplanted more effectively simply because you know where to find it at the site of the implanted pellet.

More recent still is the exciting work that the people in virus production of tumors have been doing. The elegant presentation by Dr. BIGNER yesterday was a model of its kind. The careful analysis of the dosage, the animal species and the site of injection are all terribly important in the production of these tumors. I might tell you that very recently I have learned from GABRIELE ZURHEIN in Wisconsin that she has been able to produce gliomas in hamsters with the Papova virus with which she's been working now for some time and has recently published her findings. I have also learned from KATSUO OGAWA in Okayama City and from FUSAHIRO IKUTA in Niigata, Japan, that they have isolated a type

12 adenovirus from a child's lung and with it have been able to produce gliomas in hamsters and in mice. All this work in viral production of gliomas is burgeoning.

I have to repeat once again that all this exciting work with viral production of gliomas experimentally has not yet proved that this is the cause of gliomas in man. It is perhaps suggestive and attempts are being made to associate viruses with gliomas in man, but there is no definitive proof of this as yet. It is therefore perhaps of some interest to note that in the experimental animal, the viruses are seen in the pre-cancerous lesions and never within the tumor cells themselves. And maybe we are looking with the electron microscope for viruses in the human gliomas a bit too late. In the experimental animal, certainly at the implantation site of a chemical carcinogen before a tumor develops, one can find virus particles. When the tumor develops, these particles have already disappeared. It is likely, therefore, that immunological studies may ultimately prove the viral origin of human gliomas rather than the demonstration or the cultivation of viruses from tumor sites.

Dr. RUBINSTEIN and Dr. HERMAN presented a very interesting and meticulous paper, as is the habit of these investigators, on the organ culture of human and experimental gliomas. This is a continuation of the work that some 30 years ago HARRY GREENE started when he took human tumors and transplanted them into the anterior chamber of the guinea pig's eye. The advantage of course of the experimental work of RUBINSTEIN and HERMAN is that it permits visualization and interpretation of these tumors more readily. Various studies can be done with fluorescent and immunological techniques on a neoplasm in culture that one cannot do in the eye. I have two slides that I will show you briefly to emphasize what one sees in organ culture in the eye, and then you can judge for yourselves how much more elegant Dr. RUBINSTEIN's work has been. In the first slide of a mouse's anterior chamber of the eye you can see an astrocytoma. By contrast, the second slide shows a glioblastoma multiforme. Invariably this is the appearance of each glioma when transplanted into the anterior chamber of the eye. The glioblastoma even tends to form the massive blood vessels which our radiological friends emphasize when they talk about tumor "stains" radiographically.

You have heard Dr. RUBINSTEIN emphasize the fact that his studies have revealed the differentiation of cells with the production of certain cytoplasmic specialized organelles during cultivation. We did not interpret the development of the capillary system that I illustrated in the glioblastoma transplant in the anterior chamber as evidence of differentiation. I think it was merely the growth of blood vessels which grew side by side with the tumor cells of glial type that were proliferating, and this is not an example of differentiation. We get out from cultures what we put into them. In other words, you put fibroblasts and glial cells into a culture and the fibroblasts and the glial cells proliferate. Perhaps some of the collagen and some of the basement membrane that Dr. RUBINSTEIN illustrated are not true examples of differentiation. They are possibly a production of something that had been put into the organ culture to begin with.

Dr. ARNOLD illustrated the scanning electron microscope and it reminded me so very much of the first pictures that I saw some 15 or 18 years ago with an electron microscope of the transmission variety. We looked at those pictures and wondered what they meant. We had difficulty in identifying nuclei, and the Golgi apparatus looked obscure. Only persistent efforts on the part of many investigators ultimately resulted in a tool which we now use routinely in the laboratory. There are many human tumors of the glioma variety which I now can diagnose more rapidly and more accurately with the electron microscope than I can with the ordinary conventional light microscope. I remember only 12 years or so ago that several of my pathology friends looked askance and thought I was very extravagant to invest in an electron microscope. This we use now for the trainees and the house staff; it is now a tool everybody uses and can't do without. And I predict that the time will come when workers will do what Dr. ARNOLD here is doing and will use the scanning electron microscope routinely. But it is in its infancy; it's still toddling along and frankly I couldn't differentiate, and I doubt if he could, one glioma from another at this stage.

We come to the beautiful presentation of Dr. WEINBERG's on the neuro-radiological-angiographic aspects of gliomas, and as a pathologist I can only admire the temerity that these people have who look at shadows and tell you the cytological composition of the neoplasm that they see projected. And I have to admit that in the vast majority of instances they are right. We as a group of diagnostic aides, neuropathologists, have almost outlived our usefulness in the realm of glioma diagnosis. All you have to do is get Dr. WEINBERG and he'll tell you what type of a tumor to expect. I commented yesterday briefly after his presentation and indicated that every once in a great while he will make a mistake, and maybe that's the only useful purpose that pathologists will serve in the future. This may depend more on their contributions to matters of etiology and pathogenesis and the explanation of what's going on in the vital phenomena known as carcinogenesis than on actual histological diagnosis.

We come then to Dr. HUCKMAN's presentation on computerized tomography. Well I remember two years ago sitting in an auditorium as a participant in a symposium on brain tumors and hearing Mr. AMBROSE and Dr. BULL of London showing the first films seen in the United States on the use of this tool. It's inexpensive; everyone of us ought to have one in our pocket. It only costs a little less than four hundred thousand dollars. This EMI scanner is introducing a new method of attack in localization and diagnosis of lesions. One may perhaps raise the question whether at its present stage of development the instrument is worth the cost. But I thought we saw yesterday a demonstration how effective this tool can be. Please note the new terminology that is being introduced; a numerical absorption value, CT diagnosis, EMI scan. We're going to have to learn a whole new vocabulary in addition to learning how to interpret the findings. Two years ago when Dr. BULL presented the possibilities of this instrument and I commented to a radiological friend of mine that, "This thing seems to have possibilities," he looked at me and said, "Ah, another gadget." Now he's changed his mind.

We heard some interesting remarks made all too briefly yesterday on proton radiography and the proton diagnostic accelerator. This, too, is a new field and a new trend which I think is terribly encouraging and stimulating. Results in the future are bound to be important.

My good friend PAUL BUCY made the day yesterday worthwhile for me with his presentation. It was a forthright presentation and with conviction. Not many neurosurgeons, I'm afraid, will get up on a platform and say, "Gliomas can't be cured, period." And that concerns the supratentorial tumors, of course, only. Because he knows as does everyone else, and he made it clear at the end of his comments, that the infratentorial cerebellar astrocytoma can be cured even with surgery alone. I know of at least a half dozen patients who have now gone more than 20 years without any recurrence following surgery alone when the cerebellar cyst was incised and the mural nodule removed. These patients have done well through the years. So you can cure some subtentorial gliomas even if you can't cure the supratentorial tumors.

Now a word of caution. Every modality of treatment that we have heard discussed this morning provides results on the basis frequently of, in my opinion, too few cases. A glioblastoma multiforme which was operated on by PERCIVAL BAILEY in Chicago the first year he did neurosurgery here developed a recurrence 12 years later. He performed one operation and the patient received no radiotherapy. Twelve years later the patient had a recurrence and Dr. DAVIDOFF biopsied the tumor which we confirmed as a glioblastoma. Compared with Dr. BAILEY's tumor it was still exactly the same. Without radiation at that time the patient went another two years and finally died. There were 14 years of survival with surgery as the only method of therapy. Don't count on it; glioblastomas don't do this ordinarily, but if you have a small series and an occasional patient has a long survival following chemotherapy or radiotherapy plus surgery, it doesn't prove very much. And I am not saying this, by the way, in criticism of what we heard this morning because I was as fascinated and as encouraged as anyone with the chemotherapeutic and radiotherapeutic efforts that are being made. I only want to caution about the fact that if you have just ten patients and five live a little longer than the average, it doesn't indicate that 50% will live longer in a larger series. Five out of ten isn't really 50%; you're extrapolating too far, I'm afraid, that way. It's better to say five out of ten rather than 50%.

Now, the results that Dr. HENDRICK told us about and the diagnostic problems of subtentorial tumors in childhood was a nice presentation. It interests me to note that in one year they see 50 brain tumors at the Children's Hospital, for a percentage of 2.5 of all tumors admitted to the Children's Hospital. And this reminds me of what I learned a few days ago and I would like to impart this information to those of you who haven't heard me comment on it before, namely that as of last year, there were 72 neurosurgeons certified and practicing in the Philadelphia area. When the incidence of gliomas in the Philadelphia area was calculated, it was found that one neurosurgeon would work on 1.2 cases per year. If traumatic neurosurgery and shunt operations were not done there really wouldn't be enough cases to go around on the glioma basis alone. But at the Children's Hospital in Toronto they attract patients from much of Canada, I'm sure.

And now we come to what I consider very fascinating efforts that are being made to approach the whole subject of the basis of proliferative activity in neoplasia, attempts by various chemical means to catch cells and arrest their progress and division so that they can be affected in a curative way. Some two to three years ago I visited Dr. KIRSCH's laboratory in Denver and saw him start much of this work and, knowing Dr. KIRSCH as I know him, I expected that we would make progress but he really surprised me this morning by showing the progress that he actually has made. A rational means of trying to affect cures or at least survivals with tumors is as yet purely on an experimental basis, but it is a hopeful manifestation of what is going on. And I agree fully with Dr. PAUL BUCY's statement of yesterday that the future hope for therapy in gliomas in man lies in the people who resort to basic sciences as aids in treatment.

I object to the use of terms such as "partial cure." It reminds me of the term "slightly pregnant." You are either cured or you are surviving a certain length of time, and to talk of a five year cure I think isn't really justified. I have seen many a patient with a breast carcinoma who died 12 years later. She was "cured" for 5 years, but she wasn't cured of the cancer. It's better, I think, not to talk about cures in gliomas at this stage, when we all admit that the supratentorial ones at least are almost uniformly fatal. It's better to say that a patient with a glioblastoma multiforme treated a certain way has a three year survival rather than a three year cure. I'm not saying that either Dr. KIRSCH or the other speakers have used that term. I am worried myself about the use of such terms and want to bring this in as a word of caution at this time.

Dr. WILSON I have known for many years and whenever he speaks I am enchanted. He is indefatigable. He does experiments that are overwhelming. It's hard to analyze the huge amount of material that he presents. Perhaps that's why he's so thin. His use of various combinations of drugs experimentally and in the intact animal shows imagination and frankly is the hope of the future in therapy. It is equally innovative of him to use therapeutic adjuncts such as hyperthermia in his chemotherapeutic trials with gliomas.

Now I want to issue another word of caution. We have always tended to emphasize the tumor and never the host. We always talk about the neoplasm and we show a slide of a gemistocytic astrocyte and assume that in every animal and every human body the cell will grow precisely the same way at the same rate and result in a terminal event at precisely a certain hour of a certain day of a certain month of a certain year. And this is not true. The reason we see many of the variations in survival experimentally and in man is because the host, the individual himself, modifies rates of growth and tendencies for spread and things of that sort. I'll give you an example of what I mean. From the several thousand experimental mouse tumors that I have worked with in almost 40 years of experimentation, I learned that I could produce a diminution in the size of a brain tumor by putting the mouse on a starvation diet. You starve the mouse, and he loses weight, and the brain tumor loses weight. You put him back on an adequate diet and as he starts growing, the tumor starts growing. Nutrition alone is a factor. It has almost nothing to do with the tumor cells. We have also done the following experiment which shows how the environment as well as the host plays a role in tumor growth. We have taken a classical ependymoma and put it into the chorioallantoic membrane of a chick embryo and it grew like a sarcoma, wild and infiltrating and not like the nice ependymoma it was in the mouse. At any rate, it was no longer identifiable as an ependymoma in the chick embryo. The tumor can be passed through from one chick embryo to another in subtransplants and always shows this wild growth. From the eighth serial chick transplant we've injected the tumor cells back into the mouse and low and behold we again got an ependymoma. And this we have done with other gliomas as well. So, the environment and the host are terribly important and we do not pay enough attention to them. Some of the bizarre results that we get under various types of therapy may depend upon host resistance. If you make a mouse that carries a breast carcinoma acutely sick, as has been done with an infectious agent like a trypanosome, the trypanosomes cure the tumor, almost; it seems to disappear. And then if you cure the same mouse of its trypanosomiasis, the tumor grows back again. But for a time, you think you've cured the tumor. This has been done by many people, has been repeatedly described, and it is the illness of the host that affects the tumor growth. I can't overemphasize the importance of this.

The conventional radiotherapy employed by Dr. SHELINE I think substantiates many of the results that we have been aware of. They are results, however, that have still led many neurosurgeons to seek additional methods of treatment. His was a comprehensive presentation.

Dr. CHANG's presentation dealt with unconventional radiotherapy, should I say; if the previous was conventional, his was unconventional. He's using all sorts of accessories like hyperbaric oxygen and hyperthermia. That reminds me, by the way, since he's using hyperthermia, that Dr. SUTTON reported a year ago at the American Neurological Association meeting, experiments in which he inserted hyperthermic probes into brain tumors. He thus brings parts of the tumor, not the whole animal body, up to a temperature of 42^{0}C. That stimulates, he has demonstrated, the metabolic rate locally, and the tumor cells whose metabolic rate is increased become avid for radioactive phosphorus. He injects these patients, and he has done that to a number, with radioactive phosphorus. He believes that the tumor cells that grab up the phosphorus prevent it from getting to the normal adjacent glial and neuronal tissue in the brain. The neoplastic cells would be so avid in digesting this lethal substance, that they would die of indigestion. But the trouble is that patients with glioblastoma multiforme whom he's treated have died anyway, maybe a month later than you would expect on the average.

Dr. CHANG and the other people who are working in this field have demonstrated methods of approach which I think must be encouraged. Hopefully they will ultimately solve the problem of what to do for patients with gliomas.

Now I think the time is up and you want to leave, so I thank you for listening to me.

Epilogue: Treatment of Glioma — How Far Have We Come, What Are We Doing, and Where Are We Going?

J. Hekmatpanah

It would have been an act of pretense only, if I meant to convey, by treating a broad title such as this, that I have reviewed the entire literature or even understood all that was presented in this Symposium about gliomas. Nor would I even delude myself that if I did, I could have commented fairly in a few pages. But in contrast to a court of law where every piece of evidence should be examined, in a subject matter such as that of this Symposium, it might not be improper to consult some random or selected material. For the former one has to come up with a verdict, but for the latter one might come up with a notion, or at least a personal opinion. It is in this sphere that I attempt to treat such a title, and in so doing not all the materials reviewed are referred to.

What makes MR. RICHMAN GODLEE's operation on November 25, 1884 so important is not only its novelty and the post-operative survival of the patient. The cranial operation and even the removal of brain tumors had been performed before (2), although in The Lancet (3) it was reported that the operation was performed "for the first time in the history of medicine". But by this operation it seems that GODLEE opened up the gate to a new era behind which the surgeons had been barred - pent up, impatient, and constrained. Impatient because of the demands upon them to treat the sick suffering from intracranial tumors with all their ailing symptoms. Constrained, because of the fear of unknown and devastating results which could have ensued after handling an organ that for so long had been considered to be mysterious. Although the functional localizations of the brain, the antiseptic measures, and the anesthesia, the three most important discoveries that had to be made before such an operation could be performed, were done, there still were dilemmas in all three areas. The functional localizations of the brain, conceived even some centuries B.C. by HIPPOCRATES and GALEN (2), and further proven by BROCA (5), FRITSCH and HITZIG (13), were not unanimously agreed upon. Even BROWN-SEQUARD (6), who had so elegantly worked out the symptoms of hemisection of the spinal cord, was not willing to consider similar regional functions of the brain. He was particularly against the contralateral representation of motor functions when he remarked "Since the time of Mistichelli's discovery (in 1709) of the decussation of the anterior pyramids of the medulla oblongata, it has been universally admitted that one lateral half of the brain is the centre, or contains the centres, for the voluntary movements in the half of the body on the opposite side. Very strong - I may say decisive - arguments can be arrayed against this theory". In spite of all the disputes over anesthesia, from the discovery of nitrous oxide by JOSEPH PRIESTLY in 1772 to the historic administration of ether anesthesia by MORTON for the operation done by J. C. WARREN on October 16, 1846, and all the other anesthetic discoveries later, the surgeons were still reluctant to use it (18, 21, 26). These disputes were not all on academic bases; some were based on religion and some seem to have been based on rivalry. Not infrequently the experiments on the effects of the anesthesia were carried out on the scientist himself. HORACE WELLS, an advocator of nitrous oxide for anesthesia, in 1844 received this gas for the extraction of his own tooth. But when he extracted a tooth from a patient the discovery was called a hoax because the patient screamed, although later he admitted that he had felt no pain (18). Finally DR. WELLS committed suicide with chloroform in 1848. Whether his suicide was caused by depression over the death of one of his patients during the experiments with anesthetic in 1844 (21) or whether it was caused by "addiction to chloroform" and disintegration of his memory (26), would at this time be irrelevant. But it is important to note that in those days, not too unlike today, some discoveries took a shorter time to make than the time that one had to take to prove them. The antiseptic measures advanced by LISTER (19) were just beginning to become a part of the operative procedure. The impact of the PASTEUR discoveries was recognized by few, and the dispute over "spontaneous generation" was just beginning to fade (2, 3).

GODLEE's operation, therefore, besides its novelty, was important for the simultaneous use of all the above discoveries. DR. BENNET's correct diagnosis of the site of the tumor gave strength especially to the theory of functional localization of the brain because the patient had no external evidence pointing to the lesion. As good chance had it, MACEWEN and HORSLEY were living at that time, ready to carry on the task. MACEWEN (20) had already operated on other brain lesions, including a tumor. Also he had not forgotten the death of a patient whose brain abscess he had correctly localized, but the patient's parents refused to permit the operation. Within a short time he reported his own cases of craniotomies. HORSLEY (15), with his immense knowledge of laboratory experiments on animals, became the first truly qualified neurological surgeon.

This new specialty, however, required a man like HARVEY CUSHING in order to gain the needed prestige and recognition as one of the main branches of surgery. Although the fact was established that one could easily operate on the brain, the high mortality involved was not an easy matter to accept, and the operation was done by only a few. Were it not for CUSHING's energetic mind and body, his determination, and his compassion for the sick, he could not have prided himself and his profession in 1931 by reporting the 8.7% mortality for all tumors and 11% for gliomas in his later series (8).

So far as the treatment of brain tumors is concerned, the discovery of ventriculography via ventricular puncture in 1918 and via spinal puncture in 1919 by DANDY (9,10), and also the discovery of cerebral angiography by MONIZ (22, 23) in 1927 are of no less importance than the discovery of functional localization of the brain. The significance of these valuable diagnostic tests was well conveyed by DR. FRANCIS GRANT (14) when he tried to describe what neurosurgery was in the early 1920's. DR. GRANT and DR. CLARISSA RUSSELL, having "secretly" gone over some 125 cases of brain tumors referred to DR. FRAZIER by DR. SPILLER for operation, discovered that in 33 1/3% of these cases no tumor was found. The diagnostic ability of no collaborating neurologist and neurosurgeon of today could be any better than that of these scholars were it not for all of today's ancillary diagnostic tests. In addition to negative explorations, it was not unusual for neurosurgeons of those days to operate on the wrong side or on the wrong location of the head. DR. CUSHING's mistaken operation on the pituitary region in a patient with a large "cyst" in the posterior fossa in 1910 is a classic example of false localization in those days. Nevertheless, when ventriculography was discovered, he denied his patients and himself the benefit of this discovery. DR. FRAZIER at that time refused to consider it. The story of DR. CUSHING's disapproval of ventriculography is of course well known. Even in the lecture that he felt was his "last" on the subject of brain tumors as a whole, ventriculography was barely discussed. He finally acknowledged some value for ventriculography in more precise localization in "obscure cases" (8). In regard to angiography, even DANDY had the following comment (2): "If there were not a better method this might be entertained, but since ventriculography can localize all brain tumors with a single injection, is perfectly safe and leaves no permanent deposit, there can be no place for angiography in this field. Its only function is in the disclosure of unlocalizable aneurysms of the brain and this it does very beautifully." He apparently did not realize the potential of angiography in rendering even a histological diagnosis in some brain tumors. Angiography is now used far more frequently than ventriculography for cerebral tumors.

It was in such a climate that a group of men like DR. BUCY gradually entered the scene, carrying on the flag through the yet unknown passages. They tested the theories and the techniques, they operated and followed their results, they learned and taught, they met and formed societies, until through their efforts they brought about the existing knowledge in operative neurology.

The purpose of this historical account is not to elaborate on what has just been presented. The speakers, I am sure, could have treated this subject better had the allotted time and space been longer. But it is only in the light of the historical events that one can fully appreciate the tasks of those before us and the progress which enables the neurosurgeons of today to have such low operative mortalities. We feel deeply grateful to them.

While a great many scientific and technological advances have occurred in all areas of the field of neurosurgery, if we exclude cerebellar astrocytomas and a few other low-grade gliomas of the brain, today we are able to contribute hardly more to the happiness of the patient with malignant glioma than they were in the first case, 90 years ago. This is especially true for gliomas which are biologically malignant such as glioblastomas, or locally malignant such as gliomas of the brain stem. GODLEE's patient died one month later of secondary infection (4), but today's patients die of recurrences some six to twenty months after operation, radiation, or both. Unfortunately, some 30% of all intracranial tumors would fall in this group. So far as the very low mortality and temporary relief of symptoms is concerned, we can surely be proud, but I have used the term "happiness" without which there could perhaps be very little meaning in life. During the past decade I have paid very specific attention to this aspect, and have interviewed almost all my own and some of my colleagues' patients with glioblastomas. In their follow-up, I have also interviewed their close relatives. The operations consisted of wide excisions of the tumors to secure as large an internal decompression as possible without removing the speech areas, sensory and motor cortex and their tracts, or deep gray matter. If a portion of the tumor was invading these areas, consisting of mainly tumor or necrotic cysts, the excision continued. They also received radiation after the operation. Post-operatively, almost everyone was happy. But I must say, however, that I found few patients or relatives with genuine happiness some six to nine months after the operation. The temporary happiness of the relatives was soon spoiled by the sight of a slowly deteriorating and helpless patient before them. They were often bothered by one or several feelings, such as depression, fear, death wishes, guilt, and anger. The regained alertness of the patient often seemed to have helped him only to learn what had happened to him and what was going to happen.

The results of the operation on cerebellar gliomas are of course, quite different. While in the early part of this century the operation on cerebellar tumors was considered "futile to attempt" (25), the operation could very well be curative now. This is particularly true in the case of cystic astrocytoma. The ventricular shunting of posterior fossa tumors for periods before, during, and following the operation, has proved to be a valuable means of reducing the morbidity and mortality of all posterior fossa tumors. We have had experience with this measure for ten years and in some fifty patients. The value of the procedure is beyond doubt. I find it unfortunate that the procedure is not as widely used as its benefits warrant. These benefits are not only for preoperative and operative reduction of increased intracranial pressure, thereby reducing the complications, they are also of great value for early and late post-operative hydrocephalus. The symptoms of mild chronic hydrocephalus can erroneously be taken for post-operative disturbance of cerebellar function.

The fact that a definitive treatment for glioma is not yet discovered indicates that the problem is not as simple as it had been hoped before. Not infrequently, one hears the expression of hope for "finding the cause of cancer." There is already convincing evidence to indicate that the causes are many, and provided that the cause of one type of cancer is found, we have no reason to assume that the cause will be found for the others. Nor can we assume that whenever the cause is found the treatment is guaranteed to follow. The research on gliomas must therefore have its own independent inquiry by those who have a specific interest in this cancer, who also have a general background knowledge of tumor biology. It took almost one half of a century for scholars such as TOOTH (27), RIBBERT (24), BAILEY and CUSHING (1), HORTEGA (16), KERNOHAN (17) and ZULCH (28, 29), to classify the gliomas. A correlation between the cytoarchitectural appearance of tumors with the duration of their symptoms before and the duration of the recurrence after the operation is perhaps one of the most important biologic aspects of these works. A good bulk of the time seems to have been spent on the development of a language for communication so that when a tumor was given a name the term would signify the cytogenesis and the degree of malignancy. The biologic behavior of a malignant cell as compared with that of the normal cell and its life cycle was not very much known. The experimental tumor production in animals, tissue cultures and transplants, some of which are presented in this Symposium, has helped toward understanding the behavior of the cells in addition to their appearance. Granted that the understanding of the behavior of the cancer cells does not necessarily procure a successful treatment, it might yet prove to be a prerequisite.

The recent interest of some centers in more aggressive treatment of malignant brain tumors, beyond what operation and radiation can offer, is of importance. As one can judge through what has been presented here, there is some indication for better results, but certainly not satisfactory results.

In summary, therefore, it seems that a great deal of progress has been made in recognition, classification, biology and radiologic diagnosis of gliomas. But so far as the treatment for malignant gliomas is concerned, we are only slightly ahead of where we were 90 years ago. The chemotherapeutic approach is offering some evidence for hope, but a "hope" only. There seems to be a unanimous agreement that the hope for a definitive treatment for glioma should be in the domain of "basic science." This prediction is not very daring, in that we can claim it without the fear of being proven wrong someday. For in every real art or technique we can find some trace of earlier science. If man was able to land on the moon it was because he was ready to come up with a proposal to estimate what it took, in what length of time, and with what cost, to do it. In the case of gliomas, there is no evidence that we are capable of coming up with such a proposal at the present time. Let us go back, therefore, to our hope as stated in the "Preface" of this book, to be able to learn where we stand with gliomas. It can be stated, at least for my part, that this Symposium was of real help in giving us a bird's-eye view of where we stand with gliomas and what is being done by some representatives of the front runners. But so far as where we are going, we don't even know.

REFERENCES

1. BAILEY, P., CUSHING, H.: A Classification of the Tumors of the Glioma Group on a Histo-Genetic Basis with a Correlated Study of Progress. Philadelphia, London & Montreal: J. B. Lippincott Company 1926.
2. BALLANCE, SIR CHARLES A.: A Glimpse into the History of the Surgery of the Brain. The Thomas Vicary Lecture, Royal College of Surgeons of England (Dec. 8, 1921). London: Macmillan and Co., Limited 1922.
3. BENNET, H., GODLEE, R.: Excision of a tumour from the brain. The Lancet 1884 II, 1090-1091.
4. BENNET, A. H., GODLEE, R.: Case of cerebral tumour. The surgical treatment. Medico-Chirurgical Transactions 68, 243 (1885).
5. BROCA, M. PAUL: Remarques sur le siège de la faculté du langage articulé, suivies d'une observation d'aphémie (perte de la parole). Bulletins de la Societé Anatomique de Paris 6 (Series 2) 330 (1861).
6. BROWN-SEQUARD, C. E.: The appearance of paralysis on the side of a lesion in the brain. The Lancet 1876 I, 2.
7. CUSHING, H.: The Pituitary Body and its Disorders. Clinical States Produced by Disorders of the Hypophysis Cerebri. pp. 194-197. Philadelphia & London: J. B. Lippincott Company 1912.
8. CUSHING, H.: Intracranial Tumours. Notes Upon a Series of Two Thousand Verified Cases with Surgical-Mortality Percentages Pertaining Thereto. Springfield, Illinois - Baltimore, Maryland: Charles C. Thomas, Publisher 1932.
9. DANDY, W. E.: Ventriculography following the injection of air into the cerebral ventricles. Ann. Surg, 68, 5 (1918).
10. DANDY, W. E.: Ventriculography of the brain after the injection of air into the spinal canal. Ann. Surg. 70, 397 (1919).
11. DANDY, W. E.: Surgery of the Brain. A Monograph from Vol. XII of Lewis' Practice of Surgery, p.122. Hagerstown, Maryland: W. F. Pryor Company, Inc. 1945.
12. FRANKLAND, P., FRANKLAND, MRS. P.: Pasteur. The Century Science Series. New York: The Macmillan Company 1898.
13. FRITSCH, G., HITZIG, E.: Über die elektrische erregbarkeit des grosshirns. Arch. für Anat., Physiol., und Wissenschaftliche Medicin (1870).
14. GRANT, F. C.: The surgery of Meningiomas at the base of the brain. Clinical Neurosurgery, Vol. 5, pp. 25-37 (Chapter II). R. G. Fisher, Ed. Baltimore: The Williams & Wilkins Company 1958.
15. HORSLEY, V.: A note on the means of topographical diagnosis of focal disease affecting the so-called motor region of the cerebral cortex. Am. J. Med. Sci. 93, 342 (1887).
16. HORTEGA, DEL RIO P.: Estructura y sistematisación de los gliomas y paragliomas. Arch. Espan. Oncol. 2. 411 (1932).

17. KERNOHAN, J. W.,SAYRE,G. F.: Tumors of the Central Nervous System. Atlas of Tumor Pathology, Section X, Fascicles 35 and 37. Washington, D.C.: Armed Forces Institute of Pathology 1952.
18. KEYES, T. E.: The History of Surgical Anesthesia. New York: Schuman's 1945.
19. LISTER, J.: On a new method of treating compound fracture,abscess, etc. The Lancet 1867 I, 326.
20. MACEWEN, W.: On the surgery of the brain and spinal cord. Med. News 53, 169 (1888).
21. MASSENGILL, S. E.: A Sketch of Medicine and Pharmacy and a View of its Progress by the Massengill Family from the Fifteenth to the Twentieth Century. Bristol, Tennessee: The S. E. Massengill Company 1942.
22. MONIZ, E.: L'encéphalographie artérielle, son importance dans la localization des tumeurs cérébrales. Rev. Neurol. 342, 72 (1927).
23. MONIZ, E., ALVES, A.: L'importance diagnostique de l'arteriographie de la fosse posterieure. Rev. Neurol. 40, 91 (1933).
24. RIBBERT, H.: Über das spongioblastom und das gliom. Virchows Arch. Path. Anat. 225, 195 (1918).
25. STARR, M. A.: The results of surgical treatment of brain tumors. J. Nervous Mental Disease 30, 398 (1903).
26. THORWALD, J.: The Century of the Surgeon. New York: Pantheon Books 1956.
27. TOOTH, H. H.: Some observations on the growth and survival period of intracranial tumors, based on the records of 500 cases, with special reference to the pathology of the gliomata. Brain 35, 61 (1912).
28. ZULCH, K. J.: Brain Tumors, Their Biology and Pathology. American Edition, based on the Second German Edition. Berlin, Heidelberg, New York: Springer Publishing Company, Inc. 1957.
29. ZULCH, K. J., WOOLF, A. L., Editors: Classification of brain tumours. (Report of the International Symposium at Cologne 30th August-1st September 1961). Acta Neurochir. Supplementum X. (1964).

Recent Results in Cancer Research

Sponsored by the Swiss League against Cancer. Editor in Chief: P. Rentchnick, Genève

1 Schindler, R.: Die tierische Zelle in Zellkultur. Geb. DM 22,—, US $ 9.50
2 Neuroblastomas — Biochemical Studies. Edited by C. Bohuon (Symposium). Cloth DM 22,—; US $ 9.50
3 Hueper, W. C.: Occupational and Environmental Cancers of the Respiratory System. Cloth DM 45,—, US $ 19.40
4 Goldman, L.: Laser Cancer Research. Cloth DM 22,—; US $ 9.50
5 Metcalf, D.: The Thymus. Its Role in Immune Responses, Leukaemia Development and Carcinogenesis. Cloth DM 32,—; US $ 13.80
6 Malignant Transformation by Viruses. Edited by W. H. Kirsten (Symposium). Cloth DM 43,—; US $ 18.50
7 Moertel, Ch. G.: Multiple Primary Malignant Neoplasms. Their Incidence and Significance. Cloth DM 25,—; US $ 10.80
8 New Trends in the Treatment of Cancer. Edited by L. Manuila, S. Moles, and P. Rentchnick. Cloth DM 43,—; US $ 18.50
9 Lindenmann, J., and P. A. Klein: Immunological Aspects of Viral Oncolysis. Cloth DM 25,—; US $ 10.80
10 Nelson, R. S.: Radioactive Phosphorus in the Diagnosis of Gastrointestinal Cancer. Cloth DM 21,—; US $ 9.10
11 Freeman, R. G., and J. M. Knox: Treatment of Skin Cancer. Cloth DM 21,—; US $ 9.10
12 Lynch, H. T.: Hereditary Factors in Carcinoma. Cloth DM 32,—; US $ 13.80
13 Tumours in Children. Edited by H. B. Marsden and J. K. Steward. Cloth DM 96,—; US $ 41.30
14 Odartchenko, N.: Production Cellulaire Erythropoïétique. Relié DM 38,—; US $ 16.40
15 Sokoloff, B.: Carcinoid and Serotonin.
16 Jacobs, M. L.: Malignant Lymphomas Cloth DM 32,—; US $ 13.80 and Their Management. Cloth DM 25,—; US $ 10.80
17 Normal and Malignant Cell Growth. Edited by R. J. M. Fry, M. L. Griem, and W. H. Kirsten (Symposium). Cloth DM 70,—; US $ 30.10
18 Anglesio, E.: The Treatment of Hodgkin's Disease. Cloth DM 30,—; US $ 12.90
19 Bannasch, P.: The Cytoplasm of Hepatocytes during Carcinogenesis. Electron- and Lightmicroscopical Investigations of the Nitrosomorpholine-intoxicated Rat Liver. Cloth DM 40,—; US $ 17.20
20 Rubidomycin. A new Agent against Cancer. Edited by J. Bernard, R. Paul, M. Boiron, C. Jacquillat, and R. Maral. Cloth DM 59,—; US $ 25.40
21 Scientific Basis of Cancer Chemotherapy. Edited by G. Mathé (Symposium). Cloth DM 34,—; US $ 14.70
22 Koldovský, P.: Tumor Specific Transplantation Antigen. Cloth DM 30,—; US $ 12.90
23 Fuchs, W. A., J. W. Davidson, and H. W. Fischer: Lymphography in Cancer. With contributions by G. Jantet and H. Rösler. Cloth DM 93,—; US $ 40.00
24 Hayward, J.: Hormones and Human Breast Cancer. An Account of 15 Years Study. Cloth DM 42,—; US $ 18.10
25 Roy-Burman, P.: Analogues of Nucleic Acid Components. Mechanisms of Action. Cloth DM 34,—; US $ 14.70
26 Tumors of the Liver. Edited by G. T. Pack and A. H. Islami. Cloth DM 68,—; US $ 29.30
27 Szymendera, J.: Bone Mineral Metabolism in Cancer. Cloth DM 40,—; US $ 17.20
28 Meek, E. S.: Antitumour and Antiviral Substances of Natural Origin. Cloth DM 20,—; US $ 8.60
29 Aseptic Environments and Cancer Treatment. Edited by G. Mathé (Symposium). Cloth DM 30,—; US $ 12.90
30 Advances in the Treatment of Acute (Blastic) Leukemias. Edited by G. Mathé (Symposium). Cloth DM 46,—; US $ 19.80
31 Denoix, P.: Treatment of Malignant Breast Tumors. Indications and Results. Cloth DM 58,—; US $ 25.00

32 NELSON, R. S.: Endoscopy in Gastric Cancer. Cloth DM 58,—; US $ 25.00

33 Experimental and Clinical Effects of L-Asparaginase. Edited by E. GRUNDMANN and H. F. OETTGEN (Symposium). Cloth DM 71,—; US $ 30.60

34 Chemistry and Biological Actions of 4-Nitroquinolin 1-Oxide. Edited by H. ENDO, T. ONO, and T. SUGIMURA. Cloth DM 40,—; US $ 17.20

35 PENN, I.: Malignant Tumors in Organ Transplant Recipients. Cloth DM 30,—; US $ 12.90

36 Current Concepts in the Management of Lymphoma and Leukemia. Edited by J. E. ULTMANN, M. L. GRIEM, W. H. KIRSTEN, and R. W. WISSLER (Symposium). Cloth DM 53,—; US $ 22.80

37 CHIAPPA, S., R. MUSUMECI, and C. USLENGHI: Endolymphatic Radiotherapy in Malignant Lymphomas. With contributions by G. BONADONNA, B. DAMASCELLI, G. FAVA, F. PIZZETTI, U. VERONESI. Cloth DM 53,—; US $ 22.80

38 KOLLER, P. C.: The Role om Chromosomes in Cancer Biology. Cloth DM 53,—; US $ 22.80

39 Current Problems in the Epidemiology of Cancer and Lymphomas. Edited by E. GRUNDMANN and H. TULINIUS (Symposium). Cloth DM 64,—; US $ 27.60

40 LANGLEY, F. A., A. C. CROMPTON: Epithelial Abnormalities of the Cervix Uteri. Cloth DM 64,—; US $ 27.60

41 Tumours in a Tropical Country. A Survey of Uganda (1964—1968). Edited by A. C. TEMPLETON. Cloth DM 79,—; US $ 34.00

42 Breast Cancer: A Challenging Problem. Edited by M. L. GRIEM, E. V. JENSEN, J. E. ULTMANN, and R. W. WISSLER (Symposium). Cloth DM 53,—; US $ 22.80

43 Nomenclature, Methodology and Results of Clinical Trials in Acute Leukemias. Edited by G. MATHÉ, P. POUILLART, and L. SCHWARZENBERG (Symposium). Cloth DM 64,—; US $ 27.60

44 Special Topics in Carcinogenesis. Edited by E. GRUNDMANN (Symposium). Cloth DM 58,—; US $ 25.00

45 KOLDOVSKÝ, P.: Carcinoembryonic Antigens. Cloth DM 38,—; US $ 16.40

46 Diagnosis and Therapy of Malignant Lymphoma. Edited by K. MUSSHOFF (Symposium). Cloth DM 62,—; US $ 26.70

47 Investigation and Stimulation of Immunity in Cancer Patients. Edited by G. MATHÉ and R. WEINER (Symposium). Cloth DM 96,—; US $ 41.30

48 Platinum Coordination Complexes in Cancer Chemotherapy. Edited by T. A. CONNORS and J. J. ROBERTS (Symposium). Cloth DM 68,—; US $ 29.30

49 Complications of Cancer Chemotherapy. Edited by G. MATHÉ and R. K. OLDHAM (Symposium). Cloth DM 58,—; US $ 25.00

50 Cancer Registry. Edited by E. GRUNDMANN and E. PEDERSEN (Symposium). Cloth DM 48,—; US $ 20.70

51 Gliomas. Current Concepts in Biology, Diagnosis and Therapy. Edited by J. HEKMATPANAH (Symposium). Cloth DM 54,—; US $ 23.30

Special Supplement: Biology of Amphibian Tumors. Edited by M. MIZELL. Cloth DM 106,—; US $ 45.60

In Production

52 Cytostatic Therapy — A Double-Edged Sword? Edited by E. GRUNDMANN and R. GROSS (Symposium).

Prices are subject to change without notice

MIX
Papier aus verantwortungsvollen Quellen
Paper from responsible sources
FSC® C105338

If you have any concerns about our products,
you can contact us on
ProductSafety@springernature.com

In case Publisher is established outside the EU,
the EU authorized representative is:
Springer Nature Customer Service Center GmbH
Europaplatz 3, 69115 Heidelberg, Germany

Printed by Libri Plureos GmbH
in Hamburg, Germany